Competence and Vulnerability in Biomedical Research

INTERNATIONAL LIBRARY OF ETHICS, LAW, AND THE NEW MEDICINE

VOLUME 40

For other titles published in this series, go to
www.springer.com/series/6224

Phil Bielby

Competence and Vulnerability in Biomedical Research

Springer

Phil Bielby
Law School and Institute of Applied Ethics
The University of Hull
UK

ISBN 978-1-4020-8603-8 e-ISBN 978-1-4020-8604-5

Library of Congress Control Number: 2008931007

Printed on acid-free paper

9 8 7 6 5 4 3 2 1

springer.com

Acknowledgements

This book derives from my doctoral thesis, awarded by The University of Sheffield. I am particularly grateful to my examiners, Søren Holm and Mark Taylor, for their perceptive comments and constructive criticism, which have influenced the revisions I have made in converting the thesis into a book. I must also thank Deryck Beyleveld for his insightful and patient supervision of the original thesis, which considerably enhanced the clarity of my thinking on these issues and the cogency of the arguments. More recently, the two anonymous reviewers for Springer provided useful feedback on the manuscript and suggestions for improvement. I have been fortunate to be a part of the Sheffield Institute of Biotechnological Law and Ethics and, latterly, the Institute of Applied Ethics at Hull, which have both offered vibrant inter-disciplinary research cultures in which to explore ideas. A number of colleagues and friends have provided helpful advice, comments and information at various times, especially Don Carrick, Bev Clucas, Fiona Cownie, Sam Hardy, John McMillan, Shaun Pattinson and Tony Ward. Apologies to anyone I have overlooked. Of course, none of them are responsible for any shortcomings in the text.

The University of Sheffield School of Law supported the first three years of my doctoral studies through a generous scholarship, which allowed me to commence research on this topic earlier than I otherwise might. A period of research leave from The University of Hull Law School in the first semester of the academic year 2006–07 facilitated work on the manuscript, and I am grateful to colleagues who covered my teaching and administrative responsibilities during this time. In addition, I would like to express considerable thanks to my publishing editor at Springer, Fritz Schmuhl, and to Marion Wagenaar for their efficiency and patience in waiting for the final version of the manuscript.

Chapters One and Seven include and develop material originally presented in my article 'The Conflation of Competence and Capacity in English Medical Law: a philosophical critique' (Bielby 2005b), which appeared in *Medicine, Health Care and Philosophy*, published by Springer. Chapter Two contains previously published work drawn from my article 'Equality and Vulnerability in Biomedical Research on Human Subjects' (Bielby, 2005a), and I am grateful to Steve Smith, editor of *Imprints: egalitarian theory and practice*, for permission to use this material.

Excerpts from my forthcoming book chapter, 'Towards supported decision-making in biomedical research with cognitively vulnerable adults' (Bielby, 2008), which will be published as part of *The Limits of Consent: A socio-ethical approach to human subject research in medicine* (Corrigan et al eds.), appear in Chapters Three, Six and Seven. I wish to thank the publishers, Oxford University Press, for granting permission for their use here.

On a personal note, my parents and sister have shown bounteous moral support and hospitality, particularly in the period when I was commuting between Sheffield and Hull. At home, Mappin and Flapjack have offered much feline friendship, especially during late nights working in the study. My deepest gratitude, however, is to Hannah, my wife. Throughout, her encouragement, optimism, empathy and wry humour have sustained me enormously. In the final stages, her help proved especially invaluable. The many personal sacrifices she has willingly made to allow me to finish my doctoral thesis – and then to indulge my keenness to turn it into a book – are a measure of how extraordinarily caring (and tolerant) she is. I dedicate this book to her.

28 March 2008 Phil Bielby

Contents

Main Abbreviations

FPCLC	First-person contemporaneous legal competence
GCAb	Generic Capacities of Action behaviour
GCAbrpvp	Generic Capacities of Action behaviour consistent with rpvp
PGC	Principle of Generic Consistency
PTDCJ	precautionary task or decisional competence judgment
rpvp	relation: proactively valuing purposes

Note on the Text

In the interests of textual fluency, where the gendered pronoun 'he' or 'she' would be used to denote a hypothetical human agent, the agent is always referred to as 'she'. I sometimes refer to an agent as 'it', not to dehumanize human agents, but to emphasize that there is nothing inherent the concept of an agent that suggests that the agent must be gendered or sexed, or even human.

Where discussed, I have stated the legal and professional ethics regulatory position correct as of 1 March 2008.

Introduction

One might be forgiven for thinking that decisional competence is not a central concern of biomedical research ethics. Many bioethicists working in this area, for example, have instead focused on human rights abuses in the history of medical research with human participants,[1] the arguments in support of the principle of free and informed consent to research participation[2] and the ethics of biomedical research in developing countries.[3] More recently, the emerging field of 'neuroethics' has directed interest towards enhancement technologies and the social implications of neuroscience.[4] However, the ethical basis for making judgments of decisional competence to consent to biomedical research on behalf of individuals with some form of cognitive vulnerability raises issues which touch on all of the above. It is quite surprising, therefore, that the issue has not been more widely discussed.[5]

---

[1] See Moreno (2001), Jonsen (1998, Chapter 5), Lederer (1997), Advisory Committee on Human Radiation Experiments (1996), Annas and Grodin (eds.) (1995), McNeill (1993, Chapters 1 and 2), Lifton (1988), Katz (1972), Pappworth (1967, esp. Part 1), Beecher (1966), Mitscherlich and Mielke (1962) and Fox (1959). I use 'participant' in preference to the more conventional 'subject' as I wish to avoid the connotation of passivity and subordination associated with being a subject. Not all agree with the nomenclature of research participant, however; see Grinnell (2004, esp. pp. 73–74).

[2] See Katz (2002), Berg et al. (2001, Chapter 12 and 13), Foster (2001, Chapter 8, esp. pp. 113–128), Doyal and Tobias (eds.) (2000), Fischman (2000), Wellman (1999: 133–136), King et al. (1999), Bloche (1998), Brody (1998, esp. pp. 43–48), Meisel and Kuczewski (1996), Katz (1993), Veatch (1987), Sumner (1987: 182–194), Freund (1972).

[3] See, especially, Macklin (2004), Resnik (2004) and Benatar (2001).

[4] The essays collected in Illes (2006) and Marcus (2002) offer an excellent overview of this field.

[5] Much of the literature relevant to this research question has appeared over the last decade. These include: DuBois (2007, Chapter 5), Karlawish (2007), Saks and Jeste (2006), British Medical Association and The Law Society (2004, Chapter 13), Howe et al. (2003), Jeste et al. (2003), Toner and Schwartz (2003), Wendler and Shah (2003), Appelbaum (2002), Roberts (2002), Maio (2002), Poythress (2002), Appelbaum and Grisso (2001), Berg and Appelbaum (1999), Appelbaum et al. (1999), Capron (1999), Appelbaum (1998), Elliott (1998), DeRenzo et al. (1998), Arboleda-Florez and Weisstub (1998), Weisstub and Arboleda-Florez (1998), Weisstub et al. (1998a, b), Bonnie (1997), DeRenzo (1997), Dresser (1997), Elliot (1997) and Glass (1997). Older discussions can be found in Berg (1996), Cutter and Shelp (eds.) (1991), Kopelman (1990), Morreim (1983), Keith-Spiegel (1983) and Macklin (1983, who considers treatment and research).

There have been a number of attempts to address the concept of competence *per se*. However, these attempts have tended to adopt a narrower discipline-specific approach, not usually venturing beyond philosophy or psychiatry, and engaging with an academic audience drawn from those disciplines.[6] They have also tended to cluster around a number of established topics. Where commentators have attempted to develop a multi-disciplinary account of competence to consent to medical interventions, these have tended to proceed with regard to the separate question of consent to treatment and not research.[7] Other attempts to theorize competence have considered it in relation to particular social behaviours and practices, including competence to consent to sexual intercourse, competence to initiate divorce proceedings and competence to stand trial.[8]

In the context of biomedical research, a presumption or judgment of decisional competence serves to ascribe specific cognitive and conative skills to an individual, which affirms their capability to decide whether to participate on their own behalf.[9] Knowledge of decisional competence is significant for three reasons. First, it occupies a 'gate-keeping'[10] function to identify the circumstances in which it is possible to seek contemporaneous consent from the person whom the decision will affect[11] and in which seeking such consent is ethically justified. Where there is doubt, assessing competence fulfils a determinative function concerning whether a particular person is able or unable to consent on her own behalf.

---

[6] For example, Shanteau et al. (2004), Checkland (2001), Buller (1991), Wilks (1999), Cale (1999), Wicclair (1999), Wilks (1997), Goldstein and Hogarth (1997), Wicclair (1991a, b), Brock (1991), Sternberg and Kolligian Jr. (1990), and from a jurisprudential perspective, Spaak (1994, 2003).

[7] Tan et al. (2006), Northoff (2006), Benaroyo and Widdershoven (2004), Breden and Vollmann (2004), Silver (2002), DeMarco (2002), Welie and Welie (2001), Charland (2001), Holm (2001), Vaswani (2000), Gunn et al. (1999), Grisso and Appelbaum (1998), Wear (1998, Chapter 7), White (1994), Devereux (1993), Cutter and Shelp (eds.) (1991), Buchanan and Brock (1990, Chapters 1 and 2), Faden et al. (1986, Chapter 8, esp. pp. 287–294), Culver and Gert (1982: 42–63), Tancredi (1982), Freedman (1981), Abernethy and Lundin (1980), Baumgarten (1980), Roth et al. (1977). Jonas (2007) is a recent exception insofar as she also addresses research. Zaubler et al. (1996) provide a useful annotated bibliography of research published in this area between 1977 and 1995.

[8] See Wertheimer (2003, Chapter 10), Grisso (2002), Stefan (1993) and Poythress et al. (2002, Chapter 2).

[9] This should not negate the important role of emotion in decision-making. Emotion is implicated in practical reasoning in so far as it constitutes an expression of cognition and evaluation rather than entailing a separation from it. See Damasio (1994) and Nussbaum (2001) for lucid accounts of the cognitive and evaluative dimensions of emotion respectively, and Silverman (1997) and Charland (1998a) for an application of these ideas to decisional competence to consent. For a more circumspect view of the relationship between emotion and decisional competence, see Appelbaum (1998a).

[10] A term first used in Faden et al. (1986: 274). "[C]ompetence judgments function to distinguish persons from whom consent should be solicited from those from whom consent need not or should not be solicited" (*ibid.*: 288).

[11] As opposed to proxy consent or a prospective consent provided through an advanced directive.

Second, the rationale for making judgments about decisional competence to consent to biomedical research and the extent to which decisional competence informs ascriptions of first-person contemporaneous legal competence (FPCLC)[12] depends upon the type of ethical theory which grounds judgments of competence. As Benjamin Freedman put it in a seminal article over twenty-five years ago, the ethical dimension of competence judgments mediates between their ascriptive and normative status:

> Competence is not pure description … nor pure evaluation; but, perhaps, an evaluation that stems from certain described facts coupled with a moral theory that tells us what those facts mean, how we ought to respond to those facts. A middle step between description and action is always needed, i.e., a moral theory that interprets experience and guides action.[13]

Knowledge of who is judged decisionally competent to consent and who is not makes clear whether consent from the person who is to participate in the research is sufficient for the purposes of authorising research participation or whether further authorisation (e.g. from a proxy) is required. It also reveals how values such as autonomy and paternalism, beneficence and social justice are understood and reconciled when addressing issues of competence. A few years before Freedman's article, Jeffrie Murphy presciently identified the ethical question this raises:

> [T]he vast majority of cases that confront us will be borderline – cases in that greyish area between full competence and obvious incompetence. The real problem that will face us, then, is what to do in the borderline cases. When in doubt, which way should we err – on the side of safety or on the side of liberty? It is vital that we do not adopt analyses of "incompetence" or patterns of argument that obscure the obviously moral nature of this question.[14]

My aim in this book is to provide an answer to this ethical question. I intend to develop a justificatory framework for making judgments of decisional competence to offer a contemporaneous consent to biomedical research for individuals whose decisional competence may be questionable, by applying rights-based moral reasoning informed by empirical studies in psychiatry. By 'questionable', I mean that there is enough doubt surrounding it as to suggest an assessment of competence, but not so much that their decisional competence is likely to be widely fluctuating or at the point of atrophy. My focus is therefore on individuals who could plausibly consent on their own behalf, without the need for a surrogate decision-maker or advance directive. I consider five discrete groups – older children and adolescents, adults with intellectual disabilities, adults with depression, adults with schizophrenia and adults with dementia, including Alzheimer's disease. I refer to these groups collectively as 'cognitively vulnerable' participants.[15] The argument draws upon

---

[12] I will discuss FPCLC in more detail in Chapter 1.

[13] Freedman (1981: 55).

[14] Murphy (1979a: 174).

[15] My use of cognitive vulnerability is separate from that of Beyleveld (2008), who applies it to mean the rational limitation all agents face in the understanding of morality (*ibid.* 85). Karlawish (2007) also uses the expression 'cognitively vulnerable', but in a different way to denote individuals who lack competence to make a decision rather than about whom it is questionable (2007: 599). Alternatively, Freedman (1981) refers to such individuals as being "marginally competent" (*ibid.*: 54). I do not use this term as it begs the question of where the margins of decisional competence lie.

ideas from moral philosophy, bioethics, psychiatry, philosophy of mind and legal theory and seeks to integrate relevant insights that each offers.[16] In doing so, I intend to develop a novel theory of judging decisional competence to consent to biomedical research.

In Chapter 1, I explain and distinguish five concepts of competence – agency, task, decisional, societal and legal competence – in order to understand their meaning and inter-relationship and to achieve conceptual clarity. From this, we can arrive at an analytical understanding of what we are invoking when we discuss competence and in what context. The definitions of competence we will arrive at are separate from the normative question of when to judge the presence of or to support the development of decisional competence to research in relation to cognitively vulnerable individuals, a question that a definition of competence alone cannot answer.

In Chapter 2, I discuss the nature of competence to consent[17] and why I use 'research' in preference to 'experimentation'. This leads into an analysis of human vulnerability (which could also be thought of as agent vulnerability), along with its significance to biomedical research with participants whose decisional competence to consent may be questionable. I argue that adolescents and adults with mental disorder or intellectual disabilities embody to varying degrees and in different ways a notion of heightened intrinsic vulnerability that I call cognitive vulnerability. This is comprised of factors internal to the individual that *may* affect decisional competence, such as cognitive immaturity or underdevelopment, misperception of reality or declining memory retention. These are sufficient for such individuals to have their decisional competence brought into doubt but may not be sufficient to lead to a judgment of decisional incompetence. The nature and purpose of the 'new' brain sciences and other advances in biomedical research often depends upon the participation of such individuals whose decisional competence may be *prima facie* questionable.[18]

Defining decisional competence – which I will argue is more coherent in its 'decision-relative' than 'risk-relative' form – provides a basis upon which we can

---

[16] In taking such an approach, I will align my argument with a perspective set out in an earlier work on the theory of consent: "The special commitments of medicine, law, psychiatry, philosophy, psychology, and other professions have led to competing perspectives on competence that are in many instances incompatible. Some have claimed that there is not and likely will never be a consensus definition of competence. This view is short sighted" (Faden et al. 1986: 288). This book goes further, however, as it seeks to prove that there can be a consensus theory of competence, insofar as I attempt to reach a rapprochement between the "special commitments" of relevant disciplines.

[17] Although this understanding of competence could be coherently applied to other scenarios that involve sophisticated decision-making abilities.

[18] For an overview of recent developments in the area of neuroscience, see Merkel et al. (2007, Part I), Ackerman (2006), Rees and Rose (eds.) (2004) and Andreasen (2001). On the impact of neuroscientific, psychopharmacological and genetic knowledge on medical research with adolescents and adults with mental disorder or intellectual disabilities, see: Andreasen (ed.) (2005), Hanin et al. (2005), Jellinger et al. (eds.) (2000), Rapoport (eds.) (2000), Williams et al. (eds.) (1999), Ebert and Ebmeier (eds.) (1998), Wahlström (ed.) (1998), Brunello et al. (eds.) (1997), Hall (1996) and Häfner and Wolpert (1996).

undertake normative and empirical analysis of making judgments of competence to consent to biomedical research. In Chapter 3, the moral theory applied in this book is introduced, a morally objectivistic, dialectically necessary approach developed by the North American philosopher, Alan Gewirth, known as the Principle of Generic Consistency (PGC).[19] On a meta-ethical level, the PGC commits all agents to accepting that they have moral rights for no other reason than that they are agents.[20] For the purposes of the book, I will take Gewirth to be essentially correct in his theory and will not seek to defend it from counter-arguments.[21]

After setting out the stages of the argument, I explore the value that the PGC attaches to consent and explain why the PGC supports a positive rights-claim to assistance in decision-making. When applied to the context of medical research, the PGC endorses recognition of the decisional competence of research participants and support for its development where possible. I conclude the chapter by illustrating the epistemological superiority of the PGC as a justification for moral rights and duties through a comparison with the contractarian moral theories of John Rawls and David Gauthier. I have chosen Rawls and Gauthier as comparators because, like Gewirth, they both detail the steps of their arguments carefully and seek to provide an explicit rational foundation for morality.

Setting out one's moral theory, however, is not enough to identify the scope of the moral community it presupposes. In Chapter 4, I explain how the PGC is used to identify members of the moral community by applying the PGC under conditions of metaphysical uncertainty, which Beyleveld, Brownsword and Pattinson have termed 'precautionary reasoning'. I consider how we ultimately cannot know with certainty that there are other agents apart from oneself, but that this should not lead us to act on the premise that there are no other agents than ourselves. Precautionary reasoning requires us to treat any entity (human or otherwise) as an agent where the available evidence suggests that the entity has agentive capacities. Even when it appears not to have these capacities, we still cannot assume that the entity in question is not an agent, although it will affect how paternalistically we treat that entity.[22] Thus, the recipients of our moral obligations need not merely be those who can apparently exercise rights and duties themselves. In doing so, this chapter lays the groundwork for the more specific application of precautionary judgments and gives shape to the meaning of precaution which I employ in this book. Whilst there has been discussion of the role of the 'precautionary

---

[19] Gewirth was born in 1912 and died in 2004. His most discussed work is *Reason and Morality* (Gewirth, 1978).

[20] These moral rights are subsumed under the broad categories of rights to freedom and well-being.

[21] Beyleveld (1991) has already done this extensively, to which I refer the interested reader. Later criticisms of the argument are discussed and refuted in Beyleveld and Brownsword (2001, Chapter 4) and Beyleveld (2002). Additionally, Regis (1984) and Boylan (1999) bring together several moral philosophers to debate the nature and implications of the argument, which include replies by Gewirth.

[22] Beyleveld and Pattinson (1998, 2000), developed in Beyleveld and Brownsword (2001).

principle' in bioethics, there have been no previous attempts to apply 'precautionary' moral judgments in bioethics (grounded in the PGC or otherwise) to judgments of decisional competence.[23] In this sense, the meaning of 'precaution' within the argument differs from previous usages.

In Chapter 5, I apply precautionary reasoning to the agentive, task, and decisional capabilities of members of the five cognitively vulnerable groups under consideration in this book. I draw upon psychological research, psychiatric nosologies and debates about personal identity in the philosophy of mind as a basis for making ascriptions of agency, and reflect upon the likelihood of each group being able to develop task and decisional competences. This generates a theory which I call precautionary task or decisional competence judgment (PTDCJ). PTDCJ specifies that the correct judgment to make about the task or decisional abilities of an agent is the one that has the least harmful consequences for the generic rights of the agent and those of any other agents whose generic rights may be affected compared with the effects of any other judgment that may be open in the situation.

In the context of making judgments of competence, PTDCJ requires that where the available evidence suggests that an agent can exercise a particular task or decisional competence, the agent should be presumed able to do this. Similarly, where the evidence suggests that the agent can develop this competence with appropriate support, PTDCJ requires that the agent should be presumed able to do this, and as far as possible be given this support. In both cases, the judgment cannot be made if the consequences of permitting the agent to exercise or develop this competence are worse (as measured under the PGC) than not permitting the agent to exercise or develop this competence.

Chapter 6 presents a critical survey of the empirical studies currently available on decisional competence to consent in medical and related forms of research with the five cognitively vulnerable groups under consideration in this book. Evidence derived from studies and assessments of competence to consent to medical research substantiates the claim that a considerable number of members of these groups are likely to have at least a developable decisional competence to make participation decisions on their own behalf which may be realised once offered appropriate support. When making a judgment of decisional competence under PTDCJ, the empirical evidence for decisional competence can be categorized broadly into three groups: compelling; uncompelling and inconclusive. The factors that place the decisional competence of cognitively vulnerable individuals within these three categories include psychological maturity, ability to generate preferences in relation to specific decisional contexts, level of minimally adequate cognitive functioning (including emotional integrity) and the liability of decisional competence to fluctuate over time. However, the limitations of the existing research attenuate the force of the conclusions we can draw.

[23] For an overview of the role of the precautionary principle in bioethics, see Kopelman et al. (2004) and Häyry (2005). For a criticism, see Harris and Holm (2002). Beyleveld and Brownsword (2007b) offer a Gewirthian defence and reconstruction of the principle.

Chapters 7 and 8 extend the theoretical and empirical analysis to examine decisional competence to consent to biomedical research in policy proposals, ethical guidance and medical law in England and Wales and the United States. This serves to establish to what extent decisional competence informs ascriptions of FPCLC to consent and identifies whether rules or policies exist to offer support in making decisions about research participation to adolescents, adults with mental disorder and adults with intellectual disabilities. We will gain a sense of the meaning and importance attached to decisional competence to consent as a practical principle and identify the similarities and differences with the position supported by PTDCJ.

In conclusion, by applying Gewirthian moral reasoning under precaution,[24] we can make morally defensible judgments about the decisional competence or incompetence of a potential cognitively vulnerable participant to consent on their own behalf. We can reach these by having regard to whether, given the available evidence, a judgment of decisional competence would be more harmful to the individual's agency rights than a judgment of decisional incompetence. This empirically-informed normative approach minimizes the possibility of reaching conclusions that are unethical, implausible, or inconsistent with other findings. In this book, it is not my aim to provide practical recommendations to guide the development of law and ethical codes in respect of competence to consent to biomedical research. Instead, I intend to make the case for a theoretical framework to which any future development should have regard.

At this stage, I should acknowledge that there are those who are sceptical that we actually can have recourse to a rationally grounded set of ethical principles with which to resolve dilemmas in bioethics and medical ethics.[25] In anticipation of the opponents of moral foundationalism (and to avoid becoming a hostage to fortune), I concede that the ethical argument in support of this justificatory framework may proceed merely from an acceptance of the existence of human (or agency) rights and need not necessarily rest upon acceptance of the dialectical necessity of the PGC for its force.[26] Moreover, there are alternative egalitarian rights-based moral theories which would imply, if not entail similar conclusions.[27] Yet this is, I believe, unlikely to be enough to convince the hard-line sceptic precisely why rights are indispensable independently of their enshrinement in a system of positive law.

As audacious as this may seem to some, I believe that the Principle of Generic Consistency is the most compelling answer yet offered in response to the 'authoritative question' of moral philosophy, namely, 'why should I be moral?' This does not foreclose continuing debate in moral philosophy by any means, but instead provides a robust touchstone against which we should measure the cogency of future contributions. A foundationalist, dialectically necessary account of morality

[24] Beyleveld and Pattinson (1998, 2000), Beyleveld and Brownsword (2001).

[25] See, for example, Baum (1990: 7).

[26] Beyleveld (1996).

[27] Such as Rawls (1972, 1993) or Dworkin (1978, 2000).

has the advantage of not leading to question-begging conclusions derived from premises which ultimately cannot be proven. Answering the authoritative question is a fundamental requirement in seeking to repudiate our hard-line sceptic and for considering what grounds we have for making moral judgments at all, let alone the grounds on which we should make judgments about decisional competence for research participation from cognitively vulnerable participants. It is in the belief that this question is answerable – and that Gewirth has persuasively answered it – that I have written this book.[28]

[28] Other commentators who have espoused or applied Gewirth's ideas include Phillips (1986), Beyleveld and Brownsword (1986, 2001, 2007a), Beyleveld (1991), Toddington (1993), Pluhar (1995), Brown (1998), Harris (2002), Pattinson (2002), Clucas (2004) and Spence (2006).

Chapter 1
Five Concepts of Competence

That commentators use 'competence' frequently within moral philosophy, psychiatry and legal theory is reason enough not to doubt its multi-disciplinary credentials.[29] It would be mistaken, though, to suppose that the insights of each discipline are either uncontested or incontestable. Moral philosophy, psychiatry and legal theory operate according to different discourses and utilise esoteric bodies of knowledge. For the uninitiated, these can often be highly abstruse. They typically rest upon disparate theoretical and empirical premises and reach different conclusions, sometimes from different perspectives within the same discipline. In particular, the way in which competence has been conceptualised previously has led to insights from some disciplines to be given more prominence than others, and for other disciplines to be neglected altogether. A fusion of these insights into an integrated theory of decisional competence to consent to biomedical research is no easy task. We must begin, therefore, by drawing distinctions.

Although there has been much debate about how to assess competence, relatively little attention has been paid to defining it.[30] Competence (or 'competency'[31]) is a normative ethical quality, cognitive-psychological trait and legal property and takes several forms. For the sake of clarity, I arrange these into a lexical ordering:

1. Agency competence
2. Task competence
3. Decisional competence
4. Societal competence
5. Legal competence (also known as 'legal capacity')

In the ordering, agency competence is ontologically the most fundamental whereas legal competence is the most contingent. However, all these elements are constitutive of the broader meaning of competence insofar as to negate one element would be to impoverish our explanation of another.

---

[29] See for two notable examples, Landry (1999) and Pepper-Smith et al. (1996).

[30] White (1994: 55).

[31] As it is often known in North America.

Agency Competence

Common to each form of competence is a measure of successful generic or specific human function or capability. Beauchamp describes this as:

> a single, basic, skeletal meaning that underlies various criteria of competence: '*X* is competent to do *Y*' always means '*X* has the ability to perform task *Y*' 'Competence' thus *means* 'the ability to perform a task'. This is the term's simple definition (and its logically necessary and sufficient condition).[32]

Beauchamp's description expresses competence as an ability. This is true of all human competences, as its logical presupposition is that the individual concerned possesses the appropriate qualities to perform or participate in *X*. However, competence at its definitional level does not entail that *X* should be a *task*. Task competence is a type of competence which presupposes the existence of antecedent capabilities. These antecedent capabilities are necessary for human action *at all*, namely consciousness, perception, ratiocination and volition. They are qualities common to all conatively functional human beings who we may consider competent in the most basic sense. I will call this type of competence *agency competence*. Agency competence is the necessary and sufficient condition of the development or possession of any other competence. Without this basic competence, individuals lack the qualities necessary for minimally independent human life.

Agency competence entails an ability to generate freely chosen purposes and categorically instrumentally value the necessary means to those purposes (irrespective of what those purposes might be).[33] At the most fundamental level, the pursuit of any purpose requires freedom and well-being, by which I refer both to the ability to deliberate and make choices that are expressive of one's intentions and to the possession of fundamental goods that are presupposed by and needed to sustain this, such as life, sustenance and basic knowledge.[34] This current or prospective ability for action is a necessary and sufficient condition of being an agent.[35]

Shoemaker describes agency as the "close cousin" of personhood.[36] Indeed, Carruthers has defined personhood in terms of agency.[37] However, I use agency in preference to personhood, for two reasons. Agency is the least restrictive criterion to denote the purposiveness ascribed to persons, as it denotes the capacity for freely chosen action construed as broadly as possible. On this account, all the agent is committed to valuing are its purposes and the means to attain those purposes,

---

[32] Beauchamp (1991: 50).

[33] Gewirth (1978: 22, 25–27, 44), Beyleveld (1991: xxxvi) and Beyleveld and Brownsword (2001: 73).

[34] Gewirth (1978: 30–47), see also Beyleveld (1991: 18–21).

[35] Gewirth (1978: 46 and 119–127).

[36] Shoemaker (2007).

[37] Carruthers (1986: 234), although Carruthers does so with reference to 'rational' rather than 'moral' agency (*ibid.*: 228, 232).

whatever those may be. By comparison, Harris, for example, connects personhood with the capacity to value *existence* rather than the capacity to value particular purposes.[38] Although Harris makes clear that his conception of personhood does allow a person to cease valuing their life,[39] agency tied to unspecified ends stands less in need of such qualification, as a purpose could conceivably refer to an intention to die in a context where existence is, or never has been, of value to the agent.

Second, the use of agency (at least in its Gewirthian sense) avoids semantic correspondence with being human. Correspondence usage occurs where personhood is conceived as a 'subclass' of human agency[40] or where agency is understood as partially constitutive of human personhood. Rom Harré employs this later usage, defining personhood (or 'personal being') as comprising consciousness, agency and autobiography (personal identity).[41] Insofar as human agents (the subject of this book) are conscious and capable of forming an identity, they fulfil Harré's definition of personhood. Despite this, one could infer from such an account that human beings exhaust the range of agents that exist. Such an inference would overlook the wider reach of agency. An agent theoretically could be any entity that acts for freely chosen purposes, which includes (potentially) forms of artificial intelligence and higher-order non-human animals. Agents thus comprise a very expansive natural kind. This does not mean that it is impossible to define a person in similarly broad terms, as Harris does.[42] However, the meaning of agency is less ambiguous, as it lacks the ordinary language use associated with personhood. This ordinary language use of personhood is synonymous with being human.[43] Agency is, in this sense at least, not bound up with being human in the way that personhood is.

Agency competence is not equivalent to, and nor does it require the level of cognitive function required for the exercise of advanced abilities, such as academic or emotional intelligence, abstract reasoning or practical knowledge. The possession of these specialised abilities may affect the scope of the agent's specific task and decisional competences, but not the competence of the agent *qua* agent. The threshold of agency competence is low, and as such, most human beings will appear to meet it. Human beings who appear to lack agency competence include, for example, foetuses and neonates, those in a coma or persistent vegetative state and those in the most advanced stages of dementia.[44] Their apparent lack of agency

---

[38]Harris (1985, 1999).

[39]Harris (2005b: 388–389).

[40]See Taylor (1985a, b) and Heinimaa (2000). Carruthers also identifies "distinctive human emotions and affective ties" as conditions of personhood (1986: 234), although concedes that on this account some human beings may not be persons (*ibid.*) and that it may be possible in the future to create artificial persons that are not humans (*ibid.*: 248). Such artificial entities would, according to Carruthers, have to display these distinctively human qualities to qualify as persons, however.

[41]Harré (1984). Erde (1999) offers a lucid distillation of Harré's conception of personhood.

[42]Harris (1985).

[43]Ford (2005: 80–81).

[44]See Chapter 5 for a discussion.

competence prevents any development of task or decisional competences, even though the prospective emergence or return of agency competence, and thus the potential for developing task or decisional competences, may be possible.[45] At this stage, however, I wish to emphasize that I do not see these human beings (or, for that matter, non-human animals) as being unentitled to intrinsic moral consideration. I will justify this claim when I return to the issue in Chapter 4.[46]

Task Competence

Task competence is a descriptive *and* evaluative measure that includes generic activities, such as speech or physical mobility, and highly specialised activities, such as mountaineering or performing neurosurgery. Fundamentally, a judgment of task competence serves to *describe* a level of ability shown by an individual in respect of a given task, i.e.: *A* is competent at task *B* to degree *C*. Task competence is thus a scalar quality. As the judgment of competence is task specific, it can describe a person's ability to perform or participate in an activity *P* or a range of activities P_n at the same time she is incompetent to perform or participate in other activity *Q* or range of activities Q_n. So, for example, a person may be competent to cook a meal or to speak German fluently whilst at the same time she is incompetent to play the cello or to fly a plane.

The following example illustrates the distinction between task competence and agency competence. Suppose I wish to drive from Sheffield to Leeds in order to perform in a music festival when I arrive. The mere fact that I am evincing a purpose that I wish to pursue is enough proof of my agency competence. Beyond, this, I am dependent on an array of task competences to allow me to fulfil my purposes. First, I would need to have demonstrated the competence of driving a car safely and responsibly to the satisfaction of the relevant authority. Second, I would need to have the aptitude to perform a particular musical piece on my chosen instrument(s). Third, I would probably also need to successfully engage in incidental tasks such as setting up and packing away my musical equipment at the venue, filling my car with fuel before the journey and following a map of road directions (or the guidance of my car's satellite navigation system) between the two cities. None of these task competences has any bearing on my agency; with or without them, I am still an agent.

We can also use a judgment of task competence in three comparative ways. First, a judgment of competence can have a comparative dimension in respect of the same individual over time. Here, the function of the judgment is to evaluate the current

[45] In the case of foetuses and neonates who survive to reach more advanced developmental stages or human beings who recover from a coma.

[46] The reason for my use of the qualifying term 'apparent' will, I hope, also become clear.

level of ability demonstrated by an individual, *A*, in terms of previous levels of ability that she has shown at that task under different circumstances in the past. This takes the form:

A is competent at task *B* to degree *C* in circumstances *X* at $time_2$ which is greater/lesser than the degree to which *A* previously displayed competence at task *B* at $time_1$ in different circumstances *Y*.

Second, a judgment of competence can also have a comparative dimension in respect of different persons at the same time. In this case, the judgment functions to evaluate the ability shown in terms of the level of ability at that task which another individual or group of individuals shows or has shown relative to *A*. This takes the form:

A is competent at task *B* to degree *C* which is greater/lesser than the degree to which another individual *D* is or other individuals D_n are competent at *B*.[47]

Third, a judgment of task competence may function to evaluate the task competences possessed by an individual in terms of the *probability* of the individual possessing those task competences in different circumstances. That is to say, from the fact that *A* has task competences *M* and *N* in circumstances *X*, it is possible to infer the probability of *A* having task competences *M* and *N* in different circumstances *Y*. We can also re-frame this in terms of the *probable* range of task competences that another individual or individuals possess relative to *A*. So, from the fact that *A* has task competences *M* and *N* in circumstances *X*, it is possible to infer the probability of another individual *D* or other individuals D_n having task competences *M* and *N*.

White identifies capacities and knowledge as providing the basis of task competence:

> A person is competent to perform a task, the actions of which are specified, if he knows what actions are required, knows how to perform those actions, possesses the capacities necessary to perform those actions, and, given his position, can reasonably be expected to possess both that knowledge and those capacities.[48]

White's definition is plausible insofar as she defines capacities as being physical and mental[49] and knowledge as being specific as well as generic, thus not excluding any possible task competences from consideration.[50] However, the definition appears to leave aside the meaning and significance of one's 'position' on the types of competence that one may develop. If one's 'position' is taken to denote one's level of learning or professional responsibilities rather than one's ontological position

---

[47] It would also be possible for a judgment of competence to have a comparative dimension in respect of different persons at the different times.

[48] White (1994: 47).

[49] Although the volitional component of competence means that the cognitive aspects are logically prior. See further, Culver and Gert (1982: 53–54).

[50] White (1994: 45).

as an agent, this would negate the prior question of how the competence is developed and its relative importance for fundamental human action.

Tasks at which it is possible to demonstrate competence can be categorised according to their usefulness and importance for fundamental human action. In so doing, we can observe where the absence of a task competence may seriously impinge on a person's everyday life and where it may not. For instance, the skills required for speech (e.g. co-ordination of the brain with bodily vocal articulators, enunciation) and physical mobility (e.g. walking, and bodily co-ordination) are required for many human activities for most of the time, thereby enhancing personal independence and autonomy.[51] It is empirically verifiable that the majority of adult human beings are capable of exercising the skills required for speech and mobility. Those who do not or who are no longer capable of exercising them are in clear need of additional support or assistance commensurate with the debilitating impact on their lives.[52] This debilitating impact may, of course, result from a pre-existing failing of the social environment to meet the needs of such individuals.

By contrast, fewer individuals are (or ever will be) competent to go mountaineering and fewer still to perform neurosurgery. However, it is less important for the independence and well being of those concerned that such competences are developed. Because of their application to specialist fields of activity, to lack competence to perform or undertake either activity is not going to have a detrimental impact upon the everyday life of the person who is unable to develop the competence, apart, perhaps, from the thwarting of an ambition.

Acknowledging that more individuals are task competent in respect of speech than neurosurgery may seem trite, but it is important to explain. First, the degree of understanding and skill required by speech and physical mobility is lower than that required for mountaineering or neurosurgery. Given basic knowledge about typical levels of human intelligence and motivation, it is more probable that a larger number of people are capable of speech or mobility than performing neurosurgery. Second, a precondition of possessing a task or decisional competence is that an individual should have the developable potential for exercising that competence. To be competent to go mountaineering or perform neurosurgery requires higher-level cognitive abilities, such as sophisticated judgment and critical thinking, developed later in life through a combination of prior learning and experience. Even then, they are not developed by everyone to a degree that would allow them to be deemed competent to undertake those activities supervised, unsupervised or at all. To become competent to go mountaineering or perform neurosurgery respectively requires a long and intensive process of education, training and practice. Conversely, an individual can acquire the abilities for speech or physical mobility without any formal education and with only basic levels of cognitive processing.[53]

---

[51] This is true even if we conceive of independence and autonomy separately from the more perfectionist notion of human flourishing.

[52] This claim requires a normative argument, which I develop between Chapters 3 and 5.

[53] See further Piaget (1950, repr. 2001, Chapters 4 and 5).

The latency or developable quality of task competences captures a significant part of what it means to possess them. It is not difficult to think of examples to illustrate this. One could describe an individual as having a latent task competence if he or she possessed a pitch-perfect singing voice without ever having had singing lessons, or could interpret historical events without having ever studied history. In most cases, latent task competences need development and assessment against predefined criteria before they become formally recognised, but in some cases, an informal recognition is all that is required to identify a latent competence, such as the recognition of one's singing proficiency by one's fellow members of an amateur dramatics society.

To fully flourish, developable task competences require formal or systematic training which enhances the embryonic competence. So, in the last example, the informal recognition of latent competence could incline one to develop one's competence so as to have it assessed and formally recognised, such as by taking singing lessons followed by assessment. The presence of variables that suggest a possibility, if not a certainty, that one could become competent in a certain respect indicate developable competences. For instance, not all individuals will have the competence to perform neurosurgery, but some will have a developable competence in learning science at school of such a standard that would allow them to progress to study medicine if they wished, equipping them to learn about how to perform neurosurgery if they demonstrated sufficient competence at medicine. Likewise, we can ascribe developable competences to mundane activities such as cooking. Many individuals who cannot cook have the developable potential to be able to cook following appropriate instruction. This means that we cannot presently describe such individuals as having this task competence, but that they have the *potential* to develop it should they so wish.

The potential to develop specific task competences implies a possible *prospective* exercising of those competences although they may not be possessed currently. It also suggests that an individual could develop competence at a particular activity if reasonable modifications were made to the immediate environment. For example, a severely physically disabled individual may have the cognitive abilities to learn how to cook but would not be able to develop the competence to cook in an unmodified kitchen environment. Similarly, a cognitively able adult of restricted height may not be able to develop the competence to learn how to drive a car, unless she has access to a modified vehicle.[54]

Let us return to the specialised examples of task competence given initially – mountaineering and neurosurgery – in order to draw a further distinction. The way in which we conceive of the relationships and consequences involved in

[54] Modifications to the immediate environment in order to allow as many people as possible to develop task competences is a matter of social justice that rests upon the value attached to equality of opportunity within any given society and the types of activities that justify the provision of extra support and from which the person with disabilities can benefit (e.g. the provision of learning aids for dyslexic students by a local education authority).

mountaineering compared with neurosurgery gives rise to separate moral implications. Mountaineering is an activity that has implications only for the individual who has chosen to engage in that activity. Irrespective of whether duties exist to interfere with the decision of an individual who is ostensibly incompetent to go mountaineering, achievement or failure at that activity or the risk of harm caused by the activity is directly borne only by the participant.[55] Neurosurgery, on the other hand, involves an action performed on another human being, the possible outcomes of which range from the preservation or restoration of the patient's health, serious brain damage or death. They therefore impinge upon important interests that the patient has in her psychological and bodily integrity and, indeed, life. Notwithstanding a free and informed consent of the patient to the operation, it is at least grossly negligent and at worst morally abhorrent for an individual lacking task competence in neurosurgery to be permitted to perform the intervention, irrespective of the outcome for the patient. In other words, the risk of harmful consequences generated by one individual performing an activity that has direct implications for the basic interests of other individual provides *prima facie* grounds for ensuring that individuals *incompetent* at that activity be prevented from performing or participating in it.[56] A higher threshold of competence and clear prior demonstration of proficiency is thus presupposed for activities that are (i) more complex and (ii) have a risk of direct harm to others attached.

Decisional Competence

Decisional competence follows the same conceptual structure as task competence. Common to both task competence and decisional competence is that one can possess both types of competence by degree. Just as there can be degrees of competence that fall short of the minimally acceptable standard, there can be degrees of competence that exceed this. As such, one may demonstrate it to a degree which exceeds the minimum level required to be judged competent to make the decision in question, typically expressed as a 'benchmark'.

Consider, for example, the assessment of an individual to establish whether she can be legally deemed task competent to drive a car. This requires that, at the point of assessment, all candidates have reached the same benchmark standard in that particular task. However, within that group of individuals, large disparities of ability will exist. Some will have passed the test with fewer errors than others. Some may only be able to drive a car in the situations upon which they were examined,

[55] Of course it might be possible to argue that indirect harm could be caused to the family or friends of an injured or killed mountaineer or that society is harmed through the burden placed upon health care resources if the mountaineer was to become injured, although these indirect harms are more difficult to quantify and do not seem an adequate basis for restricting human action in this way. For a discussion of the rationale of prohibiting self-inflicted harm, see Feinberg (1986).

[56] This could also be taken to mean an individual not yet proven competent at that activity.

whereas others will have the competence to drive on a motorway, off-road or on a racing track. These factors are constitutive of the competence of the individual driver. Nonetheless, the fact that all have passed the assessment shows that all have been judged to reach the sufficient and necessary standard of task competence to be legally allowed to drive. This sufficient and necessary standard can be surpassed but it cannot be fallen short of if the individual is to be considered to have reached the benchmark standard.

The same principle applies to decisional competence. Recruitment practices, for instance, commonly employ benchmark standards for decisional competence. At interview, candidates often will be given a series of activities designed to assess decision-making skills relevant to the position for which they are being considered. In these circumstances, there usually will be factors other than a demonstration of adequate decisional competence that determine the candidate's suitability for the job. Nonetheless, demonstrations of decisional competence are still operating a 'gate-keeping' function here, if only to identify those candidates who would be competent to make decisions required by the post, if they were appointed. The difference between the driving test and the decision-making assessment, however, is that the degree to which the candidate surpasses the minimal standard of competence required to meet the benchmark will usually be taken into account, especially if there are more candidates than positions available. This is because it is an indication of how well the candidate is likely to perform in the post, and thus of interest to the employer, whereas it is irrelevant to passing the driving test.[57]

Therefore, we can say of all tasks and decisions that involve a single determination of competence that they employ a minimal criterion of competence. This operates as a benchmark at which level the necessary skills and abilities must *at least* be possessed. We can express benchmark measures of assessing decisional competence in the following terms:

In order for person *A* to be deemed competent to make decisions of type *X*, abilities *E* must be possessed to a necessary and sufficient level. Abilities *E* will depend upon the nature of *X*, and may well be possessed to such a degree that exceeds the requirements for *X*. The relevance of the degree to which *A* may exceed the necessary and sufficient level required to be deemed competent to make decisions of type *X* will depend upon the purpose of the test and may well be irrelevant to the judgment of competence.

Two observations can be made of the relationship between decisional competence and benchmarks tests. First, the description of how tests for decision-making operate as a benchmark is separate from the evaluation of the correctness of the normative premises upon which those tests are based (e.g. whether the benchmark of decisional competence for recruitment to a social work position is set at a high

[57] At least in the UK. However, for all drivers that pass, there will be a record of the number of faults that were observed during the test, although these make no material difference to passing the test – one cannot pass by degree.

enough level).[58] Second, the relationship between decisional competence and benchmarks tests questions the belief that a definition of competence and the grounds on which it is tested are distinct.[59] Indeed, we have seen that in defining decisional competence as a quality that may be displayed by degree, an individual could display decisional competence that exceeds any benchmark standard that a test designed to assess that competence may employ.

Approaches to conceptualising decisional competence have tended to follow one of two dominant theoretical perspectives. The distinction between these perspectives turns upon the definitional aspects of a decision and the position that risk assumes within the decision-making context.

Risk-Relative (Asymmetrical) Competence

Common to risk-relative theories of competence is that the *degree of risk* attached to the consequence of each choice for the decision-maker within a given decision-making situation determines the level of decisional competence required.[60] Risk-relative theories of competence have been formulated principally in relation to decision-making about medical treatment, and as such, the nature of the risk discussed is one which is posed to the decision-maker only and not to others (e.g. the kind of risks involved in deciding whether or not to have chemotherapy rather than those involved in a police officer deciding whether or not to drive fast down a busy city street in response to an emergency call).[61]

[58] It is apt to explain this in light of Beauchamp's view that: "[Competence] is inherently normative in the way it is used to establish the abilities and level of abilities ... [t]hus it is a mistake to infer that empirical judgments of psychological competence are free of prior evaluative commitments. The reverse is true: they are inescapably value-laden" (1991: 53). I agree with Beauchamp that the basis of competence judgments is intrinsically normative. However, it is possible to state the mode of operation of a competence benchmark descriptively (e.g. insofar as it functions to establish whether person A has reached benchmark H for decisional competence W), in order to serve as an analytical tool to deepen the normative critique of the values expressed by the benchmark.

[59] A belief endorsed by Becky Cox White: "A definition serves a theoretical function – it tells us what we mean by competence. The capacities that define competence specify the criteria for being a competent person. To be competent is to have the relevant capacities. ... Tests serve a practical function – they are tools that identify the presence or absence of capacities, and determine whether particular persons have the appropriate abilities" (1994: 54–55).

[60] Robertson's interesting analogy reflects aspects of this approach: "Competency is a filter or screen that channels our thinking by limiting the alternatives and factors to be considered. Like a command function of a computer, it opens up new issues, although unlike a computer, it is not totally neutral about the decisions and questions that follow. The final question can be reached only after leaving the narrow domain of competency and confronting the value choices between a patient-centred or other-directed approach that are presented" (1991: 144).

[61] It would be perfectly probable in theory, however, to accommodate within a risk-related theory of decisional competence risks posed to others as well as the decision-maker. However, it is less clear whether one should assume the decision-maker to be incompetent to make a choice if the risk it posed to others is disputable.

In *Deciding for Others*, Allen Buchanan and Dan Brock claim, "just because a patient is competent to consent to a treatment, it does *not* follow that the patient is competent to refuse it, and vice-versa".[62] They illustrate this by observing that a "lumbar puncture for presumed meningitis" requires a "low/minimal" level of competence, whereas refusing "surgery for simple appendectomy" requires a "high/maximal" level.[63] In other words, the greater the risk posed to the decision-maker by the choice made, the higher the standard of decisional competence should be.

For Buchanan and Brock, balancing self-determination and the welfare of the decision maker lies at the core of their own theory of competence.[64] In their view, the choice of a decision-maker may be legitimately overridden in circumstances where a relevant authority – such as a physician or researcher – believes that the choice (i) impinges on the decision-maker's well-being and (ii) the choice reached is one that the decision maker would be unlikely to make when fully mindful of her values, irrespective of whether the decision-maker actually believes her welfare to be most satisfactorily secured by this decision. In a rejoinder to a critique of their theory, Brock restates this point succinctly:

> Persons have a self-determination interest in making significant decisions about their lives, including important medical treatment decisions, for themselves and according to their own values. But they also have an interest in having their well-being protected from serious harms that would result from their choices when their decisionmaking is substantially impaired.[65]

The validity of this claim rests upon agreement as to the sufficiency of decisional impairment such as to justify overriding the outcome of a decision or making a determination of decisional incompetence. If one understands 'substantially impaired' decision-making to refer to persons who are almost certainly incapable of making their own decisions in relation to a particular matter, then there can be no disputing Brock's claim. However, this does not capture the nature of the ethical issue at stake here. The ethical issue arises where the choice itself may precipitate a reappraisal of the person's decisional competence – where competence is *questionable* rather than palpably absent. For the choice to trigger a competence reassessment, it must either run contrary to the person's well-being or be at odds with their established values. Under Buchanan and Brock's account, substantially impaired decision-making amounts to a failure to account for these two variables by the decision-maker, rather than a complete inability to decide at all. The difficulty here is drawing a distinction between instances where the decision-maker's 'true' values are distorted and may be legitimately substituted by a surrogate and where the individual's values have undergone a genuine transformation in the

---

[62] Buchanan and Brock (1990: 51–52).

[63] Buchanan and Brock (1990: 53).

[64] Buchanan and Brock (1990: 29). They describe this as a decision-relative theory in *Deciding for Others*. However, in a paper published shortly afterwards, Brock (1991) concedes its risk-related quality.

[65] Brock (1991: 106).

recent past that might be speciously attributed to the effects of the condition which she is experiencing.

James Drane, another exponent of a risk-related approach, considers that a standard of decisional competence should vary in accordance with the dangerousness and irrationality of the choice:

> [F]or those … decisions that are very dangerous, and run counter to both professional and public rationality … competence … requires an ability on the part of the decision maker to appreciate what he or she is doing. Appreciation requires the highest degree of understanding … To be competent to make apparently irrational and very dangerous choices, the patient must appreciate the implications of the medical information for his or her life.[66]

Invoking a notion of 'professional and public rationality' is tendentious in the absence of an explanation of what such rationality is and why it is to be preferred over other forms of rationality or modes of understanding.[67] It is unclear whether a contravention of this rationality would occur at any time when the relevant authority (in this case the doctor) has her view challenged or rejected by the decision-maker (the patient), regardless of how dangerous the decision-maker perceived the decision to be. Following Drane's risk-related theory of competence could therefore allow a form of strong paternalism to enter into the process of competence assessment, negating the decisional competence that the decision-maker may actually possess. The permissibility or otherwise of such strong paternalism can only be resolved by expressly grounding a theory of competence in moral philosophy.

More recently, Ian Wilks has proposed a more extensive risk-related theory of competence, known as "asymmetrical competence".[68] This shares affinities with Buchanan and Brock's account but it employs a more clearly delineated scale of risk. Wilks suggests that an individual may be competent to choose one option but not another in a particular instance of decision-making. To illustrate, Wilks draws an analogy between asymmetrical competence and Pascal's Wager[69]:

> Just as, according to the proponents of the risk standard, what you are competent to do/not to do can depend upon the risked consequences of doing it/not doing it, so, according to Pascal, what you ought to affirm/deny can depend upon the risked consequences of affirming/denying it. Hence … competence is not independent of the consequences under risk, but rather must be determined in part according to those consequences.[70]

---

[66] Drane (1985: 20).

[67] On the problems involved in appealing to reasons to justify the superiority of one moral theory over another, see Boylan (2000, Chapter 8).

[68] Wilks (1997 and 1999).

[69] An argument proposed by the seventeenth-century French philosopher Blaise Pascal (1670) which suggests belief in God is rational, in the absence of any contrary evidence. Pascal argues given the anticipated benefit of belief in God is considerably greater than disbelief, if one believes and this transpires to be correct, then one enjoys eternity in Heaven. Accordingly, if one believes and this belief ultimately is disproved, one has lost comparatively little, other than the pleasure that may have resulted from living a hedonistic life. However, if one disbelieves and this disbelief is false, eternal damnation awaits. Therefore, if we are to value our purposes and fate, rationality dictates that we ought to subscribe to belief in God. For a critique of this argument, see Beyleveld and Brownsword (2007b).

[70] Wilks (1997: 423–424).

Wilks' theory rests upon two contentious assumptions. First, to what are we to have appeal in order to judge the value of risk to the decision-maker, when the decision-maker will – if competent – be better placed than anyone to make that decision herself? Take, as a well-known example, the choice of the decisionally competent terminally ill person to refuse a proposed life-sustaining course of medical treatment. Desire for an end to life does not constitute irrationality on the part of the patient simply because of the magnitude and irreversibility of what would be lost if the patient refuses treatment. Of course, there remain good grounds for ensuring that the patient's reasoning is informed, internally coherent, consistent and not subject to external pressure.[71] However, if there are no grounds to question the patient's competence to decide, it is hard to see any persuasive grounds for declaring that the patient is decisionally incompetent for the reason that the choice they reached is not the least or less risky option available.

Second, Wilks uses the analogy with Pascal's Wager somewhat loosely. Wilks's theory of asymmetrical competence presupposes that individuals are competent at a particular decision if they choose in some ways but not in others. Pascal's exhortation is supposedly a reason for individuals of their own free will to believe in God and not a licence for someone else to deem one incapable of making a decision about atheism or agnosticism if one chooses not to believe in God. In Wilks's theory, a judgment of competence to decide is made by someone else other than the individual. Thus, someone else's standards of rationality are being imputed upon the decision-maker. In Pascal's Wager, the competence of an individual to decide is not an issue. In suggesting that belief in God is more rational than disbelief, Pascal was not suggesting that disbelief disqualifies one from making the decision *at all*. Therefore, the analogy Wilks draws is erroneous.

An external standard of rationality is imputed upon the decision-maker in risk-relative and asymmetrical theories of competence through the standard of the 'reasonable decision-maker'. The relevant authority is likely to deem the decision-maker incompetent to decide if she fails to satisfy this standard.[72] Two consequences follow from this. First, such a standard entrenches power relationships between the decision-maker and assessor, with the freedom of the decision-maker to choose being to some extent subject to control by the assessor. Such control may indeed be warranted if the patient is incapable of reaching *any* choice in respect of that decision,

---

[71] Beauchamp and Childress argue that there are circumstances where is it permissible to pressure patients or research participants to "change their beliefs or process information differently" (2001: 91). Such pressure amounts to persuasion rather than coercion, however, and does not *necessarily* mean that patients or research participants *should* be judged incompetent if they refuse to change their beliefs or process information differently. It also does not mean that we should reach a judgment of decisional incompetence if a patient or research participant decides one way rather than another.

[72] Roth et al. similarly argue, "The patient who fails to make a decision that is roughly congruent with the decision that a 'reasonable' person in like circumstances would make is viewed as incompetent. This test is probably used more often than might be admitted by both physicians and courts" (1977: 281).

but it is far more difficult to justify where it could lead to an individual being considered perfectly capable of making a choice if deciding one way, but incapable of making the choice if deciding the other.

This gives rise to the second consequence. Risk-relative theories of competence contend that the consequences of one option of a single decision may be graver or more severe than another option, which fits well with intuitive knowledge about the array of options that any given decision scenario can pose. Nevertheless, it is not clear why this risk should attach to the consequence of a particular *choice* resulting from a decision and not be inherent to the nature of the decision-making *scenario* itself. Using an example of consent to treatment, Wilks argues that to be competent to say yes but be incompetent to say no is "as if someone says, 'You can either say yes or no. If you say yes we will immediately comply. If you say no we will have to discuss the matter further, and we may comply or we may not' ".[73]

Here lies the problem. Not to comply with an individual's choice (particularly where that person's choice only has consequences for herself) implies that the individual is not sufficiently informed or informable to *deliberate* between options and reach a choice. However, the presence of options is necessary if we are to conceive of the situation as a involving a decision *at all*. The *Oxford Dictionary of Psychology* defines decision-making as "The act or process of choosing a preferred option or course of action *from a set of alternatives* [emphasis added]. It precedes and underpins almost all deliberate or voluntary behaviour."[74] For a theory of decisional competence to postulate that an individual is competent to say yes, but incompetent to say no in respect of a single decision thus incurs a contradiction that strikes at the very heart of what the decision-making process involves.[75] It suggests that the individual cannot weigh the benefits and burdens of possible outcomes necessary even for a competent 'yes' response. The 'yes' issued in the absence of such evaluative mental processes is mere acquiescence and not the product of choice. Such an individual is therefore *in*competent to make this decision *at all*, rather than competent to decide one way but not the other.[76] Asymmetrical decisional competence thus commits the same error of "confusing compliance with competence" that Wilks levels at unjustified paternalism.[77]

By way of illustration, consider the following hypothetical scenario. A young man with learning difficulties suffering from mild schizophrenia is admitted to

---

[73] Wilks (1999: 158).

[74] Colman (2001: 187).

[75] This is particularly problematic for asymmetrical competence theory if we recall Buchanan and Brock's argument that "an adequate standard of competence will focus not on the content of the patient's decision but on the *process* of the reasoning that leads up to that decision" (1990: 50), an argument which Wilks cites approvingly (1997: 414).

[76] Culver and Gert (1990: 620) also share this view.

[77] Wilks (1997: 414). Cale's response to Wilks is broadly sympathetic to the critique presented here: "While the risks related to a decision might be grounds for taking more care in assessing a person's competence, they should not provide grounds for increasing the standards by which a person's competence is assessed" (1999: 148).

hospital after a minor episode of self-harm. Whilst recovering, and in a mentally lucid state, he is approached by a clinical researcher, who informs him of a research project being currently undertaken in the hospital to determine genetic propensity of the condition. The researcher informs him that participation in the research project incorporates two stages, each of which requires consent. The first involves the taking of blood and saliva samples for subsequent analysis. The second involves CT scanning of the brain, which contains a minimal risk of an allergic reaction to the iodine-based contrast dye. After being presented with this information, the patient orally agrees to participation in both stages of the research and signs a consent form for the first stage with the assurance that a consent form in relation to the second stage will be offered to him shortly afterwards.

In the intervening period, the researcher consults the patient's medical notes to understand more about the nature and extent of the patient's learning difficulties. The researcher reflects upon this knowledge and reappraises her view of the decisional competence of the patient to consent. She accepts the consent of the patient to the first stage of the research project, but does not accept the consent of the patient to the second on the belief that whilst the patient may have the appropriate standard of decisional competence to refuse this research, he does not have the competence to consent to it. From the perspective of a risk-relative theory of competence, this is an entirely appropriate action to take. The researcher, mindful of the condition and recent behaviour of the patient, is inclined to view the patient as capable of consenting to research activity that poses no or negligible risk, but incapable of providing consent to research activity that involves a minimal level of risk. This is as a direct result, in the judgment of the researcher, of the patient not being fully able to comprehend the nature, purpose and effect of the brain scan.[78]

However, unanswered questions remain. Did the researcher take the lucidity of the patient satisfactorily into account? After all, the patient displayed ostensible signs of comprehension of the proposed research. Should this have inclined the researcher to maintain her initial judgment of decisional competence in this case? In retracting her previous judgment, it would seem that that the researcher is doubting the patient's ability to make a decision about the second stage of the research at all, rather than doubting his ability to consent to it. Now, it may well be that there is something particular to biomedical research, especially that which involves participants with mental illness or intellectual disabilities that generates wariness about supposing a threshold of decisional competence to consent that is no higher than for refusal within a single decisional scenario. However, such wariness derives from separate normative factors about a wish to ensure protection for vulnerable groups, rather than deriving from the logical structure of the decision itself. The proper place for consideration of risk in making judgments of

[78] Adopting the criteria devised by Thorpe J in the English legal case of *Re C (adult: refusal of medical treatment)* [1994] 1 All ER 819 at p. 824 to determine the decisional competence of a patient to consent to medical treatment.

decisional competence is at the level of the decision-making scenario, in knowledge of all the choice options it offers, rather than at the level of the particular choice made.

Decision-Relative Competence

Decision-relative theories of decisional competence hinge upon the notion of 'decision-specificity', which means that they are conceptually closer to task competence than risk-relative theories of decisional competence.[79] Common to decision-relative theories of decisional competence are two claims. The first is that an individual can be competent in respect of some decisions and not others at the same time. The second is that if an individual is competent in respect of a particular decision, it follows that she must be competent to make any choice which that decision allows, on the grounds that the presence of options is required for a decision-making scenario to exist, and that to deliberate between options is a necessary condition of possessing decisional competence. Let us consider the implications of decision-relative theories of decisional competence in more detail.

Mark Wicclair, a proponent of the decision-relative approach, argues that any given instance of decision-making will typically require a different level of skills and abilities from another unrelated instance:

> The relative skills and abilities vary according to the specific decision, and a standard of decision-making capacity therefore should be decision-or-task related. It is likely, for example, that there are significant differences between the cognitive skills and capacities that are required to make a reasoned decision concerning life-extending medical treatment, on the one hand, and the cognitive skills and capacities that are required to make sound financial investments, on the other hand.[80]

To claim that significant differences exist between the skills and capacities required to make a decision about life-sustaining medical treatment and those required to make financial decisions is not to deny areas of overlap between these skills and capacities but to deny that decisional competence in one entails decisional competence in the other. It is perfectly possible that a person in possession of the requisite skills to make such a decision about life-sustaining treatment (either currently or prospectively) will not possess adequate numerical, predictive or economic skills to make decisions about financial investments.

---

[79]Culver and Gert explain this in the following way: "Specific-task competence is distinguished from risk-related theories of competence in so far as they collectively define competence as a measure of the *internal* abilities of a person. Decision-relative competence is an application of task specific competence in the decision-making sphere. The two most compelling definitions of decision-relative competence to have emerged in the last fifteen years both emphasise foremost the abilities and capacities of a person to make a decision. Competence, therefore, is considered to be an attribute of personhood" (1990: 638).

[80]Wicclair (1993: 11).

Although decision-making about life-sustaining treatment and financial investments both require informed rational deliberation, the first decision clearly involves something that is far more fundamental to one's existence and basic interests than the second. Thus, the level of understanding and appreciation required in order to be competent to decide about life-sustaining treatment is commensurate with the specific demands of the decision. These include understanding the nature of the disease, appreciating the medical prognosis and weighing up the risks and benefits of having a burdensome course of medical treatment against the consequences of not having this treatment.

We can express the decision specificity of decisional competence in the following terms:

Individual *A* is competent to make decisions of type *X* but not of type *Y* due to some property in type *X* decisions that the individual can satisfy and some property in type *Y* decisions that the individual cannot satisfy.

Where the general properties of a decision to be made comprise:

1. The content of the available options
2. The relative ease or complexity with which deliberative reasoning can lead to the selection of one particular option and
3. The impact of the available options on oneself and/or others

In addition to Wicclair, Thomas Grisso and Paul Appelbaum propose decision specific criteria for decisional competence, based upon constitutive elements of decision-making common to psychiatry, bioethics and medical law.[81] When we draw together the criteria of Wicclair as well as that of Grisso and Appelbaum, the result is a five-fold 'ideal-type' definition of decisional competence:

(i) The capacity of the person to understand
(ii) The capacity of the person to reason and deliberate
(iii) The ability of the person to communicate
(iv) The capacity of the person to possess values and goals
(v) The ability of the person to recognise options, and to appreciate the significance and meaning of different options[82]

(i) and (v) differ, as understanding in (i) refers to the nature of what is proposed (e.g. catheterisation, taking of a blood sample for research) whereas (v) refers to the

---

[81] Grisso and Appelbaum (1998: 31–32).

[82] Grisso and Appelbaum (1998: 31–32), Wicclair (1991: 91). These criteria are also similar in nature and scope to Beauchamp's definition, who argues that decisional competence requires: (i) understanding and communication of the relevant information; (ii) the weighing of risks and benefits; and (iii) making a decision in the light of such knowledge and of relatively stable values (1991: 58–59). One might be inclined to view the presence of similarities between definitions of decisional competence as evidence of an emerging consensus towards a core of settled meaning. However, such a conclusion could be a hasty one to reach given that it is not merely a case of specifying the requirements for decisional competence but also specifying how we should be interpret those requirements.

options that follow from deciding in a certain way (e.g. consenting to/refusing the proposed intervention and recognising what may follow from selecting either one of those options). The third requirement, communication, is necessary in order to operationalize any theory of decisional competence. It is of course theoretically possible that an individual could be competent to decide in the absence of the ability to communicate, although in such circumstances, it would be impossible to ascertain her decisional competence. The provision of reliable evidence to support an ascription of decisional competence lies at the core of its moral and legal credibility, and the medium through which one collects this evidence is communication. As such, provided one construes communication as widely as possible, it has a legitimate place in a definition of decisional competence.[83] Taken together, then, this 'ideal-type' concept of decisional competence has five general constitutive abilities: understanding, reasoning, deliberation, communication and capacity to form voluntary purposes that are of value to the agent.[84]

Whether an individual is competent to make a *particular* decision depends upon the level of understanding required by that decision and the extent to which an individual can reason and deliberate in order to reach a choice about that decision. This explains why some individuals are competent to make basic practical decisions (such as whether or not to see a doctor if one has a pain in one's back) but not competent to make more abstract decisions (such as whether evolutionary psychology offers a plausible account of human nature). The five general constitutive abilities of decisional competence are common to both types of decision, but they are expressed and applied relative to the specific decision in question. On a decision-relative account, the substance of the entire decisional context determines whether the individual is competent to make *any choice it allows* given the risks presented by this context, unlike risk-relative theories, which determine competence with regard to the level of risk which attaches to *the specific choice made*. As such, every decisional scenario – rather than each choice – requires a different level of the five general constitutive abilities of decisional competence, depending on how innately complex, risky or demanding it is.[85] We can express this in the following way:

The *necessary* and *sufficient* reason for being competent to make decisions of type *X* but not of type *Y* is the level of possession of abilities *E*. The extent to which abilities *E* are required is determined by the nature of decision *X*, but at the most

[83] See further, Beyleveld and Brownsword (2007a: 96, 100–101).

[84] Here I broadly agree with the analysis offered by Welie and Welie (2001).

[85] It should be apparent from this that decision-relative theories of competence do take risk as seriously as risk-relative theories. What is distinctive is that, on a decision-relative account, the relevance of risk is engaged at the level of the decision, not at the level of the choice made. Since both risk-relative and decision-relative approaches to decisional competence acknowledge risk equally, risk-relative theories could be more accurately conceived of as 'choice risk-relative' theories whereas decision-relative theories alternatively could be described as 'decisional risk-relative' theories. I will continue to employ the terms 'risk-relative' and 'decision-relative' in this book, however, to minimize any terminological confusion.

basic level are those powers of understanding, reasoning, deliberation, communication and voluntary purpose formation that constitute the basis of *any* competent decision-making.

In this decision-specific formulation of decision-relative competence (as in the ideal-type definition), there is a separation of the notions "competence to" and "competence in"[86]. Wicclair concentrates upon qualities inherent to the individual when engaging in the process of decision-making, rather than invoking external variables relative to the environment in which the individual reaches a decision, or the values of the relevant authority making a judgment of decisional competence. Whilst these environmental variables do matter, they are relevant only in the sense that the decision-maker can comprehend and appreciate them. If she cannot, then she is not competent to make the decision *at all*, rather than competent to make one choice offered by that decision but not another. This avoids conflating an external judgment of environmental risk (irrelevant to the existence of decisional competence) with the decision-maker's understanding and assessment of that risk (relevant to the existence of decisional competence).

To illustrate this, let us return to the earlier example of the schizophrenic patient. We cannot treat the probability of the CT scan giving rise to an allergic reaction to the contrast dye as integral to the competence of the patient to decide whether he wishes to participate in the research process. It is a risk external to the patient, about which he will need to demonstrate understanding and appreciation in order to be judged competent. In this example, of course, it would be difficult to make changes to the procedure in order to minimize the risks involved. Where, in different cases, changes to the level of environmental risk are possible, they may alter the nature of the decision to be made, but will not affect the individual's pre-existing level of decisional competence. In other words, the risks involved in the research activity do not determine the decision-maker's competence to consent to participation, but instead constitute the nature of the decision to be made at which the individual may or may not be competent to decide.

Additionally, a decision-maker may display aberrant incompetence in reaching a choice without warranting a judgment of decisional incompetence. Beauchamp observes that a person may happen to perform an act incompetently, even if she possesses the competence not to do so.[87] We can apply this to particular instances of decision-making. For example, a person may be competent to manage her financial affairs, yet on one particular occasion invests a significant amount of money in shares in a company about which there is speculation it may be on the verge of collapse.[88] Assuming that the investment decision was a 'one-off' and affected no one other than herself, then irrespective of her reasons for doing this (she may not, of course, have any reasons), we cannot say that on the basis of this decision alone the individual has ceased to remain competent to manage her financial affairs (although we might question their decisional competence to invest on the stock market). In removing choice risk sensitivity from the decisional competence determination,

---

[86] See Beauchamp (1991: 56–59) for discussion of this distinction.

[87] Beauchamp (1991: 57).

[88] Beauchamp (1991: 57–58) also uses a related example.

and instead applying a notion of decisional risk sensitivity, decision-relative theories more readily accept occasional mistakes and lapses of reasoning. As such, they do not reach judgments of decisional incompetence lightly.

From a decision-relative perspective, there is also an affinity between *developable* decisional competence and *potential* decisional competence. An individual could be currently unable to make a decision due to lack of information but may be perfectly competent to make the decision once the information has been supplied to her. For example, an individual considering what subject to study at university may be incompetent to take the decision before having gleaned any information on the topic or critically reflected on her preferences, yet become competent when having considered what subjects are available at which institutions and what skills and abilities they require. The provision of information does not make a difference to the person's reasoning skills but it gives these skills something with which to reason – much the same way that an engine is incapable of powering a car in the absence of fuel. Just as the fuel serves to ignite the engine, so in the same way, the information 'ignites' the competence. In biomedical research, for example, a potential participant may possess sufficient powers of reasoning as to make a competent decision as to whether or not to participate, but is not able to make a decision about participation until such time as she is provided with the information needed to exercise those powers in respect of the decision to be made.[89]

In summary, theories of decision-relative decisional competence offer a more conceptually convincing account of decisional competence than do risk-relative or asymmetrical theories of decisional competence. Decision-relative theories are no less concerned with risk, but the risk-sensitivity of decision-relative theories operates at the level of the entire decisional scenario rather than at the level of a particular choice. This maintains logical coherence with what is meant by making a decision. In determining the level of competence required to make a decision by reference to the chosen outcome, risk-relative theories of decisional competence confuse what is needed to be able to make the decision *at all* with what is necessary to be competent to decide one way or the other. In doing so, they offer a problematic view of decision-making that decision-relative theories of competence, with their emphasis upon the abilities required to make a specific decision (and not to select one options over others), do not share.

Societal Competence

Societal competence refers to the display of a sufficient range of task and decisional competences for an individual to interact independently within that society or community in such a way that does not jeopardize her own or others well-being.[90]

---

[89]Beauchamp refers to this as an instance of "a perfectly competent person who cannot competently decide in the circumstances" (1991: 57).

[90]This draws upon the definition offered by Beyleveld and Brownsword (2007a: 110).

An individual's approach to societal competence is a matter of degree (for example, an older child will display more of the task and decisional competences required for societal competence than an infant), and there will usually exist a threshold standard which individuals must satisfy in order to be officially recognised as societally competent within a given society.[91]

In one sense, a judgment of an agent's societal competence will be relative to the variables which her socio-economic environment presents. It follows that the threshold at which an authority will deem an agent to possess enough relevant task and decisional competences to be judged societally competent will vary between societies.[92] In another sense, there is a common factor underlying all judgments of this type. Any form of successful interaction entails a minimally adequate level of core knowledge, communicative and inter-personal skills that living as part of a society presupposes. Whether an individual can feed, clothe and care for herself and refrain from behaviours that would violate the physical and psychological integrity of others will typically inform a judgment of societal competence. To claim that 'X is societally competent', therefore, is to recognise that the individual concerned is dispositionally equipped to successfully and independently engage with other human beings within her immediate environment. This judgment alone generally cannot account for how societally competent an individual is above this minimally adequate level, for which the language of specific task or decisional competences is more appropriate. Similarly, a judgment of societal competence cannot rest upon putative measures of social status, such popularity or wealth, as these do not bear upon the dispositional elements of successful engagement with others.

A judgment of societal incompetence, on the other hand, usually follows from display of an insufficient range of task and decisional competences for successful social interaction, or is ascribed to an individual in the absence of the potential to develop these competences. Societal incompetence in this sense is a descriptive explanation of the ways in which an individual lacks relevant task and decisional competences for successful independent interaction and, like a judgment of societal competence, is predicated upon the contingent organisation of a particular society. It is important to remember, though, that individuals who are deemed to be societally incompetent in whatever society should not be assumed to lack agency competence. As the approach to societal competence is a matter of degree, such individuals are almost always capable of valuing purposes and, furthermore, are usually capable of developing a limited range of task and decisional competences. These are often rudimentary, but in some cases may be highly specialised and developed, such as arithmetical or artistic skills.[93] Nonetheless, the effect of a

---

[91] For example, by being of a certain age or displaying certain task or decisional competences to the satisfaction of an assessor.

[92] Beyleveld and Brownsword (2007a: 110).

[93] Examples of societally incompetent individuals who may exhibit sophisticated task or decisional competences include some persons suffering from autism or schizophrenia.

judgment of societal incompetence can be highly stigmatizing.[94] It may also give rise to a self-fulfilling prophecy in that once a judgment of societal incompetence attaches to an individual, others who interact with her may (wrongly) have difficulty appreciating that the individual is still capable of forming her own plans, values and preferences.

Reaching a judgment of societal competence or incompetence is not a straightforward process. Indeed, it can often be highly problematic. Even in societies with a broad consensus over what constitutes successful interaction, it is extremely difficult to propose a calculus of skills or traits whose absence would definitively inform such a judgment. For reasons of administrative expedience, authorities will often acknowledge societal competence by reference to a legally derived age threshold of adulthood rather than based on case-by-case judgments. Under such arrangements, many individuals will display societal competence in advance of reaching this age and other individuals will fall short of societal competence after reaching this age. Consequently, some societally competent individuals will not have this status legally conferred on them as they have not yet reached the threshold age, whereas others who have surpassed this threshold but whose difficulties in everyday living have gone unnoticed (or been dismissed) by the relevant authorities may be legally assumed to be societally competent yet in fact lack this ability. The value-laden dimension of 'independent interaction' is also a source of tension, particularly when those with the power to ascribe judgments of societal competence to individuals (irrespective of whether this is on a threshold or case-by-case basis) may operate with clandestine value judgments which serve to perpetuate the disempowerment and social exclusion of certain groups.

To identify what follows from a judgment of societal competence or incompetence is not tantamount to an endorsement of the organisation of any particular society. Nor is it to endorse a value judgment about the individual concerned. We can make judgments of societal competence or incompetence in both substantively just and unjust societies, and in many cases, these judgments will differ markedly between the two. Echoing Daniel Wikler, how inclusive our societal organisation is amounts to an issue of distributive justice.[95] It is perfectly possible that the grounds on which a judgment of societal incompetence is made serve to indicate that the society is organised along substantively unjust lines. This would occur, for instance, where a judgment of societal incompetence reveals that the nature of the social organisation fails to accommodate the needs of people with physical or intellectual disabilities, unduly medicalizes psychological suffering or privileges the interests of persons with exceptionally high intelligence. Therefore, one should not infer from the fact that an individual is judged to be societally incompetent in one society that the individual could not be societally competent in *any* society, or that the individual is societally incompetent in a morally relevant way. Indeed, the conferral

---

[94] Spicker (1990) and Thornicroft (2006) explore this idea in relation to adults with intellectual disabilities and mental disorder respectively.

[95] Wikler (1979: 377).

of a judgment of societal incompetence in a particular case may be illustrative of a moral failing of that society.

Legal Competence

Legal competence – or legal capacity – is in its essential form the exercise of a legally recognised power.[96] Legal competence is *permissive* – it serves to empower a person to be legally authorised to perform or participate in a given activity.[97] The premise of legal competence is either *presumptive* (such as reaching eighteen years of age in respect of being legally competent to vote in the UK) or *demonstrable* (such as driving a car unsupervised after having passed a test) that signifies the individual is capable at that activity or in making a specific decision. The possible types of legal competence are wide. They extend to include duties assumed by virtue of one's occupation or responsibilities (e.g. doctor, teacher, parent, carer) or vested in an inanimate body, such as an institution (e.g. Parliament) or other body (e.g. a corporation).[98]

The legal presumption of competence is heavily value laden. As Eastman observes, "the law most obviously defines models of man in relation to mental capacity and responsibility".[99] In essence, legal competence upholds the value of individual self-determination such that to enshrine the power to express a choice is to protect the power to express a choice. This value epitomises contemporary rights-based liberal political and legal thought.[100] It follows that an individual must display compelling evidence for task or decisional incompetence to trigger a reassessment of competence. An example of this is where a legal rule requires a driver convicted of dangerous driving to take the driving test again. This is also true when one declares an individual incompetent to perform a task or to make a decision. An illustration of this is an adult's loss of the right to make decisions about medical treatment of a certain type on her own behalf, where she fails to display decisional competence or displays it unsatisfactorily.

Conventional understanding views legal competence as a single concept, yet within this there are three distinct meanings that are qualitatively different. These manifestations do not alter the essence of legal competence as a legal power but

---

[96] I will use the terms 'legal competence' and 'legal capacity' interchangeably.

[97] With the exception of Hohfeld's concept of a legal power (1919), there is no evidence of academic discussion of legal competence until the 1940s. The earliest known such discussion is Green (1941).

[98] For a discussion, see Spaak (1994).

[99] Eastman (1992: 161).

[100] Evidenced in different ways in the works of (amongst others): Rawls (1972, rev. ed. 1999), Nozick (1974), Dworkin (1978), Gewirth (1978), Feinberg (1980), Thompson (1990) and Waldron (1993). See Plant (1991) for an excellent general discussion. For a specific discussion in relation to bioethical issues, see Charlesworth (1993, Chapter 2).

distinguish what factors inform the power and who is the exerciser of that power. The first sense of legal competence defines competence in terms of a power vested in an individual by law to *make decisions affecting herself* in respect of a specific activity. We can express this as follows:

First Sense of Legal Competence

Person *A* has legal competence in relation to activity *B* by virtue of having attained legal threshold *C*.

Here, being person *A* and having attained legal threshold *C* are together the necessary and sufficient condition for having legal competence in relation to *B*. The legal threshold could be presumptive or demonstrative. For example, in English law, one gains legal capacity to marry and to consent to sexual intercourse by virtue of reaching a threshold age, which is presumptive. These activities require legal competence to be held *and* exercised by the participating individuals involved, as opposed to a third party. In other words, if I wish to marry or consent to sexual intercourse (presuming the consent of the other party), I must (i) have legal capacity to do this (by being of or above the requisite age) and (ii) exercise this capacity myself at the time I wish to make the decision. I call this 'first-person contemporaneous legal competence' (FPCLC). The presumption of legal competence, or its conferral following successful demonstration of the relevant activity is not absolute and the presence of manifest decisional incompetence can rebut this.

There is no necessary conceptual connection between legal competence and decisional competence, however.[101] Standards of legal competence may reflect insights from definitions of decisional and task competence to provide grounds for FPCLC. However, legal competence can be held by individuals who are themselves decisionally incompetent to make the decision or perform the task at the time the decision needs to be made or the task performed, but who have previously delegated the decision-making authority to another person. This person has the legal power to make decisions or perform tasks of a specified nature on their behalf. Legal competence therefore extends beyond denoting the decisional abilities of an individual at the

[101] In the US, medical professionals commonly define *mental* capacity as an expression of decisional competence while competence is a "legal construct" (Berg et al., 2001: 95–96; Schneider and Bramstedt, 2006). The reverse is true in England and Wales, where not only does the dualist terminology exist but also capacity is often used in an unqualified form, which leads to inconsistency and a lack of clarity over which sense of legal capacity (or whether legal or mental capacity) is being invoked. Alternatively, in parts of continental Europe, legal competence is distinguished from decisional competence. Elsewhere, I argue (Bielby, 2005b) that this terminological complexity has given rise to frequent occasions where academic commentators and judges have conflated legal capacity and decisional competence.

time the decision is to be made.[102] The effect of this is to draw a conceptual distinction at least in part between decisional/task competence and legal competence. Beauchamp notes:

> Legal competence, by contrast to psychological competence, has to do with legal capacity … as a category distinct from psychological capacity. Some persons, such as precocious minors, may have psychological ability, but not legal 'capacity'. Some persons may have legal capacity without psychological capacity. Despite the contrast, however, legal competence generally builds on psychological competence, and adds an explicit, new evaluative dimension different from the evaluation involved in selecting abilities or tests of psychological incompetence. To say that someone is legally competent is to say that no-one is justified in authorising interventions in … the person's affairs or in acting on the person's behalf.[103]

Beauchamp is right to claim that legal competence can exist in the absence of decisional competence and the presence of psychological evidence for decisional competence is no guarantee that it will be recognised officially through an ascription of legal competence. This is different from saying that not all instances of legal competence depend upon a prior assessment of decisional competence – presumptive legal competence illustrates this. In all cases, legal competence serves an enabling function when the law allows person *A* to make decisions of type *X* or participate in activities of type *Y*.

It is not necessary, however, that the person for whom the legal authorisation is designed to benefit must always exercise legal competence herself. Legal competence or capacity can take the form of a *delegable* power exercised by a third party nominated by the beneficiary to act in the interests of the beneficiary after the loss of decisional competence. This comprises the second sense of legal competence.

The transferability of legal competence is possible provided the following conditions are satisfied:

1. The beneficiary can nominate a willing surrogate.
2. The decision or task must be able to be made or performed by another under the authority of the beneficiary.
3. The decision to be taken or action to be to performed must be something that is still of relevance to the beneficiary after the onset of decisional or task incompetence.
4. The surrogate is competent to make any decisions or perform any tasks such as may be required.

Typically, this will require:

---

[102] Legal decisional competence could feasibly allow the waiver of decision-making responsibility in relation to a particular decision altogether. Decision-making of that type may be manifest through a "conscious decision not to be involved in making … implicit and deliberate decisions" (Dekkers, 2001: 185). This would be captured by a patient's exhortation to her doctor, "I trust you to make any further decisions about this course of treatment as you see fit without discussing them with me". It is unlikely that in offering such a waiver the doctor would question the patient's decisional competence, however.

[103] Beauchamp (1991: 68).

1. A legal rule which permits a surrogate appointed by a person *before* the onset of her decisional incompetence to make decisions on behalf of the person *after* the onset of her decisional incompetence and specifies the spheres of activity in which surrogate decision-making is possible.
2. Legal authorization made by a person before the onset of her decisional incompetence consenting to a particular individual assuming the role of surrogate in respect of these types of decision-making scenarios.

We can understand legal competence in this sense as a prospectively delegable legal power. Here, the person for whom the decision-making is to benefit does not exercise the legal competence to make decisions of the type specified by the legal rule, as any grounds for presuming or demonstrating decisional competence that could inform FPCLC dissipate after the onset of decisional incompetence. Instead, the legal competence to make these decisions is *exercised* by the surrogate *on behalf of* the person for whom the decision-making is to benefit, once the person concerned no longer has the decisional competence to make decisions of this type for herself.

Delegable legal competence has parallels with other forms of legal authorisation. For example, it would be perfectly possible to authorise a willing individual to drive one's car on one's behalf, provided that the surrogate driver was competent to drive cars in general and was insured to drive this particular car. Where legal competence to make a decision also involves a task, and the individual is still competent to make the decision and not perform the task, the decision may still be taken by the individual but the task may be vested in a proxy where the task is not connected to the decision, such as voting on someone else's behalf. This is because the decision-making element only – not the task element – is the *necessary and sufficient condition* of the legal competence. This explains why, on reaching the age of majority, an individual with severe physical disabilities would nevertheless be competent to vote and why individuals are not declared competent to vote as soon as they are physically old enough to be able to visit a voting booth.

We can draw a further distinction between an individual's decision to appoint a surrogate before the onset of decisional incompetence, which requires FPCLC in order to authorise the prospective powers of the surrogate, and the surrogate's decisions taken on behalf of the individual after the onset of her decisional incompetence, the legal validity of which derives from the delegated legal competence which becomes operative once the beneficiary ceases to be decisionally competent. The operation of the delegated legal competence means that the beneficiary has not lost legal competence altogether. Legal competence continues to *reside* in the individual who is decisionally incompetent at the time the decision must be made in respect of that task, except she no longer makes it contemporaneously in the first-person. This is because, as we observed, the surrogate is exercising legal competence on behalf of the beneficiary, and because the surrogate has been vested with this power as a result of a decision made by the individual whilst she still had FPCLC to do so. We can express this as follows.

Second Sense of Legal Competence

Person *A* has legal competence in relation to activity *B* after the onset of *A*'s decisional incompetence by virtue of having transferred legal competence to make decisions in relation to *B* to person *R* before the onset of *A*'s decisional incompetence (where the transfer becomes effective as soon as possible after the onset of *A*'s decisional incompetence).

Here, being person *A*, who has transferred legal competence to *R* in respect of *B* (where *B* is a matter over which *A* previously had FPCLC and over which *A* is, in the event of decisional incompetence, legally entitled to delegate the legal competence to a surrogate), is the sufficient and necessary condition of legal competence.

It is important to note the crucial difference between '*has* legal competence' in the first sense of legal competence and '*can exercise* legal competence' in the second sense of legal competence.[104] In the first case, legal competence is both possessed and exercised by and on behalf of the same person. In the second case, legal competence is possessed by the incompetent person but exercised through her nominated surrogate.

There is, of course, a time in our lives where we cannot yet exercise FPCLC but the state still vests us with legal rights. In these cases where the beneficiary is too young to be able to exercise legal competence on her own behalf, legal competence is also exercisable by individuals deemed in law to be an appropriate surrogate to exercise the power, such as the parents or legal carer of an infant or young child. I call this 'fiduciary' legal competence. We can express this form of legal competence in the following way.

Third Sense of Legal Competence

Person *A* has legal competence in relation to activity *B* due to it being exercised by *R* during the period of life before *A* becomes legally competent in the first sense (i.e. before *A* is ascribed FPCLC) in relation to *B*.

This form of legal competence is similar to the second sense insofar as it involves a surrogate exercising legal competence on behalf of the beneficiary. However, it

[104] Article 12 of the United Nations *Convention on the Rights of Persons with Disabilities* (2006) appears to distinguish between the possession and exercise of legal capacity in a different way. This defines the possession of legal capacity as the ascription of legal personhood (i.e. recognition as a legal person) and the exercise of legal capacity as the benefit conferred by the fundamental rights which attach to legal personhood. Spaak (1994, 2003) also draws a distinction but uses it in another way to describe the hypothetical possibility of changing a legal relationship (having competence) and the actual performance of the act which changes the legal relationship (exercising competence).

differs in two ways. First, the beneficiary could not have previously expressed a wish that the surrogate decision-maker have the power they do due to the beneficiary's immaturity. In this sense, it cannot emanate from FPCLC. Second, the function of surrogate decision-making here is to hold in trust the legal competence for the child with a view to the child assuming FPCLC by such time when she is either presumed in law to have FPCLC or when she has demonstrated sufficient intelligence and maturity to make decisions in relation to such matters. As such, the stewardship rationale of fiduciary legal competence differs from that of many cases of surrogate decision-making for decisionally incompetent adults, where the need for surrogate decision-making will persist throughout the rest of beneficiary's life due to the irrevocable cause of her decisional incompetence, such as dementia or coma.

White believes that the adoption of any legal standard of competence is inherently problematic. In a section of her monograph on competence headed 'Why the Law is No Help', she argues that a definition of legal competence cannot and should not proceed without the involvement of specialists in the field in which such a definition is required:

> Suppose, however, that the law decided to construct a definition of competence to resolve future hard cases. Legal scholars would still quite likely consult the experts, that is, the health professionals. In fact, if the law failed to consult medicine, medicine would – and should – insist on being involved. Any attempt to construct a definition without information from the group who knows the most about it and will be largely responsible for its implementation would be ill advised.[105]

Nevertheless, White concludes that law ultimately cannot provide such a definition,[106] largely because there is no settled view amongst health care professionals about the criteria for an operational definition of decisional competence.[107] But even if we accept that criteria for competence are contested, this does not mean that law cannot enshrine a coherent and ethically defensible definition. Given the inherent multi-disciplinary nature of competence, White is correct that any attempt to develop a legal definition of competence without appropriate consultation would be doomed to failure. However, as White appears to be referring to a legislative definition here (especially in light of the references to 'construct' and 'consult'),[108] the possibility of this is diminished. Not only medical, but psychological and philosophical correlates are more likely to inform a legislative definition of competence than they are a judicial definition, since there will be greater scope and time available for consultation of academic and professional expertise along with other interested parties, such as support and advocacy groups.[109] Moreover, this process of

[105] White (1994: 11).

[106] White (1994: 12).

[107] White (1994: 11).

[108] Additionally, White appears to dismiss the role of the courts in determining competence earlier in this section (1994: 10).

[109] It is also probable that a judicial definition of competence (in common law) would be liable to unpredictable modification or even repeal in subsequent case law.

consultation would, if undertaken properly, give rise to a definition that would seek to integrate and reconcile the insights of various disciplinary and professional perspectives.[110] As Vanderpool observes, the "competency of definitions of competency" requires a comprehensiveness that is dependent upon different disciplinary perspectives informing and, to some extent, transforming each other.[111] Whilst a legislative definition may not be able to systematise fully the intricate definitional elements of different types of competence or incompetence, this does not imply that a legislative definition cannot endorse a skeletal task or decision-relative definition. On these grounds, I believe that a legal definition of competence is possible, but its ambitions should be to give regulatory shape to the concept rather than to define it exhaustively.

A second controversy attaches to legal definitions of competence. In order to maintain certainty and consistency, definitions of legal competence that apply to FPCLC usually require a threshold or 'cut-off point'. This is often achieved through a presumptive standard determined by age. Individuals above this age are presumed legally competent to make decisions of type *X* on their own behalf[112]; individuals below this age are not. Beauchamp and Childress argue that these pragmatic guidelines which determine legal competence are distinct from criteria for decisional competence:

> We also need to distinguish two senses of *standard of competence*. In one sense, *criteria* of competence are at stake - that is, the conditions under which a person is competent. In a second sense, *standard of competence* refers to the *pragmatic guidelines* we use to determine competence. For example, a mature teenager could be competent to decide about a kidney transplant (satisfying criteria of competence) but could also be legally incompetent by virtue of age (failing pragmatic guidelines).[113]

In principle, it may seem fairer to suggest that where some individuals appear to develop the necessary decisional competence at an age below the legal threshold of presumptive competence for making decisions of that type, they should be assessed on a case-by-case basis. The realism of this argument depends upon the nature of the decision to be assessed and the number of individuals who would be subject to assessment. For instance, it is more difficult to assess each mature and intelligent fifteen year-old to determine whether she should be given the vote before reaching eighteen, than it is to assess whether each of these fifteen year-olds has developed a sufficient understanding and intelligence as to know what is involved in her medical treatment.[114] This is because having the vote involves making decisions that have direct implications for others in society whereas consent to medical treatment

[110] See n. 16 above.

[111] Vanderpool (1991: 209).

[112] Although this is usually a rebuttable presumption.

[113] Beauchamp and Childress (2001: 107, n. 37).

[114] The test devised by Lord Scarman in the English case of *Gillick v West Norfolk and Wisbech Area Health Authority* [1985] 3 All ER 402.

only has direct implications for the person consenting.[115] For this reason, assessing voting decisional competence would almost certainly require more time and effort than assessing medical treatment decisional competence.

To formulate a legal definition of competence that is framed in a sufficiently intelligible way for use in legal reasoning yet at the same time is also informed by criteria of competence that are capable of interpretation by competence assessors – doctors, psychiatrists and psychologists – requires semantic clarity.[116] However, the application of any kind of rule (legal or otherwise) risks giving rise to a substantive injustice if the values of certainty and consistency are pursued at the expense of fairness.[117] To avoid the possibility of such injustice, a normative ethical theory is required to explain the conditions under which we should make a judgment of decisional competence and incompetence, and the circumstances in which it is ethically justifiable to deny someone the legal power to make decisions of a specific type on their own behalf.[118] On this basis, an attempt to provide a legal standard of decisional competence is more likely to remain faithful to criteria for decisional competence relative to a particular decision-making context (such as biomedical research), whilst simultaneously recognising the need to retain some kind of threshold to avoid administrative unworkability.[119]

---

[115] This should be not taken to underestimate the effect that medical decision-making by young people can have on those who love and care for them, particularly where the young person's health may be adversely affected if their decision to refuse important treatment was to be respected.

[116] Beauchamp and Childress attempt to minimize semantic confusion resulting from a *standard of competence* possessing two concurrent meanings by using the term "only to mean a criterion for determining competence" (2001: 107, n. 37). Culver and Gert (1982: 55–56) similarly seek to distinguish between the two although they do so in relation to decisional incompetence and legal incompetence. They are a little optimistic, perhaps, in claiming that a judgment of legal incompetence always depends upon a prior assessment of decisional competence.

[117] Which could otherwise be expressed as a tension between substantive and formal justice. For a discussion see, Lyons (1993).

[118] I use the term 'power' in a Hohfeldian sense.

[119] Even if decisional competence could in some cases be determined on a case-by-case basis

Chapter 2
Consent, Vulnerability and Research

Competence to Give Informed Consent

Competence to give informed consent is a form of decisional competence. Not all decision-making involves consent but every instance of consent involves decision-making. It follows that in every instance of consent there has to be an exercise of decisional competence. Decisional competence is thus a prerequisite of informed consent.[120]

Faden et al. identify two meanings of informed consent.[121] The first, which they call 'sense$_1$' is "an autonomous action by a subject or patient that authorizes a professional either to involve the subject in research or to initiate a medical plan for the patient (or both)".[122] An individual gives sense$_1$ informed consent provided that she has "substantial understanding", is not subject to the control of others, has intentionality and gives her authorization to the health care professional concerned.[123] It is similar to definitions that other commentators have offered, in particular, White.[124] 'Sense$_2$' informed consent is "legally or institutionally *effective*

[120] McConnell (2000: 70).

[121] Faden et al. (1986: 276). They claim that their definition is a purely logical one that sets out a conceptual analysis of informed consent without introducing normative variables. This is problematic, in that informed consent *qua* effective authorisation intrinsically upholds the normative value of individual autonomy.

[122] Faden et al. (1986: 278).

[123] Faden et al. (1986: 278).

[124] White (1994: 50) defines competence to consent in the following terms: "A person is competent for the task of giving a free and informed consent if (1) he is generally informable and cognitively capable of performing the actions involved in making a decision, (2) he knows that the decision-making requires these tasks, (3) he knows how to perform these tasks, and (4) given his situation we can reasonably expect him to be able to make decisions". In the absence of a criterion for the legal effectiveness of that consent, White's definition of competence to consent is similar to sense$_1$ competence offered by Faden et al. However, White's definition appears not to distinguish decisional from competence task competence. It is true that the available choices in a particular decision-making instance *may* require task competence at a particular activity but decisional

[original emphasis]... authorization from a patient or subject... obtained through procedures that satisfy the rules and requirements defining a specific institutional practice in health care or in research".[125]

$Sense_1$ and $sense_2$ consent correlate respectively to non-legal prerequisites of informed consent and legal or institutional rules of recognition for informed consent to be legally valid.[126] $Sense_2$ consent must be grounded in $sense_1$ consent criteria in order to be coherent,[127] especially if the function of $sense_2$ consent is to confer legal or institutional status on a particular normative conception of consent. However, at a conceptual level, it is possible that a legal rule or principle of an ethical code can introduce normative standards of their own, quite apart from those suggested by the normative standards of $sense_1$ autonomous authorisation. Such a disparity allows a dichotomy to emerge between the level (such as age) at which decisional competence to consent may be presumed to exist on theoretical and empirical grounds and the threshold at which the law officially recognises this decisional competence.

For example, until recently, the age of consent to homosexual sexual intercourse in England was eighteen years of age – two years higher than that for heterosexual sexual intercourse. Given that there is no difference in the level of decisional competence required for consensual sexual intercourse irrespective of whether it is heterosexual or homosexual in nature,[128] then it is clear that there were different normative values expressed by the legal standard for consent to homosexual intercourse as opposed to heterosexual intercourse. This legal rule which regulated consent was not an expression of the decisional competence required in order to make that decision, but a reflection of normative assumptions (about the morality of homosexual sexual intercourse in early adulthood) quite unrelated to the requirements of decisional competence. If we analyse the legal rule in terms of $sense_2$ consent, we see that it insufficiently explains how other values apart from those involved in judging decisional competence can influence the level at which a presumptive standard of decisional competence to consent is set in law.

Faden et al.'s two senses of informed consent are consistent with first-person contemporaneous legal competence (FPCLC). However, since they are couched in terms of providing consent in the first person (i.e. the individual who is to be

competence to offer consent *does not entail* task competence at a particular activity. For example, I can be decisionally competent to choose whether it is preferable for me to travel from Sheffield to Hull by train or by car without having the task competence to drive either a train or a car – I could envisage making the journey as a rail passenger or by taxi.

[125] Faden et al. (1986: 280).

[126] Faden et al. (1986: 277–278).

[127] Faden et al. (1986: 294).

[128] The decision whether or not to have sexual intercourse with a consenting partner requires the same level of decisional competence irrespective of whether the sexual partner is of the same or different gender. This is because the gender of one's partner is not relevant – factually or morally – to the level of decisional competence required to make the decision.

affected by the decision is the one who consents), neither $sense_1$ nor $sense_2$ consent describes the nature and context of proxy consent, which corresponds to the second and third senses of legal competence outlined in the last chapter. This should not imply, as Abrams suggests, that proxy consent is a conceptually incoherent use of consent:

> If one accepts that in order for consent to be meaningful it must be given by the individual in question rather than from a third party, indeed it is questionable, at least on a theoretical level, whether 'proxy consent' is actually consent at all.[129]

Abrams's questioning as to whether proxy consent is actually informed consent derives from a confusion over of what type of competence is grounding the consent. Abrams appears to assume that consent by definition must proceed from the first person. This is mistaken. A necessary condition of a legally valid consent is legal competence. However, as we have seen, to be legally competent to consent does not require decisional competence to be exercised through FPCLC. A surrogate decision-maker – the proxy – can be legally empowered to take the decision on the decisionally incompetent person's behalf provided it meets the criteria of delegable legal competence. In this circumstance, the decisional competence rests with the proxy.

McConnell is therefore right to conclude that if a patient or research participant is decisionally incompetent, she cannot decide on her own behalf and as a result cannot waive her right to informed consent.[130] However, that does not mean that the patient or research participant cannot exercise legal competence by virtue of having a nominated surrogate appointed to take decisions on her behalf. Thus, decisional incompetence does not entail legal incompetence. Nevertheless, as consent is being exercised on behalf of someone else rather than *by* person A *for* person A, it cannot be waived.

In decision-making about medicine and healthcare, a failure to ascribe FPCLC has significant repercussions. To be denied control over one's physical and mental health through having decisions taken for oneself in respect of these matters involves others exerting an enormous amount of influence over one's life.[131] The legal ascription of FPCLC in relation to healthcare decisions to those who display evidence of relevant decisional competences serves to empower individuals capable

---

[129] (1977: 123). More recently, Wrigley (2007) has sought to question the moral justification (rather than the conceptual coherence) of proxy consent.

[130] "the waiver of that ... right [informed consent] is relevant only so long as the patient is competent. If the patient is not competent, he cannot decide; and giving him information about the risks and alternatives serves no purpose. So there is no sense in which an incompetent patient can be said intelligently to have waived his right to informed consent" McConnell (2000: 70).

[131] White and Denise (1991) argue that an individual capable of understanding but acting upon deluded premises is equally in need of paternalistic protection as someone who is completely incapable of understanding their actions. This is a view I share, but we must be careful to specify first what we mean by 'deluded'. For a discussion, see Musalek (2003, esp. pp. 155–157).

of making such decisions. This is so even if an individual subsequently asks her physician to make certain health care decisions for her.[132]

Robertson claims that individuals who are incompetent to consent are entitled to respect for their personhood, although do not require respect for their choices.[133] But without first identifying the circumstances in which we are morally permitted to withhold respect from choices, we cannot be clear where the 'trigger point' for this lies. In its absence, we risk mistaking the ability to exercise decisional competence on the basis of compelling evidence for a simulacrum of decisional competence in someone who is very likely incompetent to make that decision.[134] Implicit in Robertson's claim is that individuals who are apparently incompetent to make a decision may also be capable of evincing a choice or preference in respect of that decision. Of course, it may be the case that a decisionally incompetent individual is still capable of forming a basic view on the matter to be decided upon. However, in order to ground a reliable judgment of decisional incompetence, an individual has to demonstrate incapability at undertaking the reasoning, deliberation and appreciation necessary to reach a choice. That someone might still be capable of choosing and expressing preferences is to raise evidence that they might not be incompetent to make that decision after all and, if possible, should have their decisional competence reassessed. It also suggests that the original evidence that supported a determination of decisional incompetence is no longer accurate or is incomplete. Establishing a strategy for minimizing error – or at least the moral consequences of this error – in judging decisional competence is, therefore, crucial.[135]

It is possible, alternatively, that Robertson is referring to legal competence, where it is factually possible for a person to be decisionally competent and to be legally incompetent (in the sense of not having FPCLC) simultaneously.[136] In these circumstances, the individual concerned would not have their choices and

---

[132] I see nothing that would be inconsistent with the concept of legal competence in allowing decisionally competent individuals to ask a relevant authority (such as a physician) to make decisions that they are decisionally competent to make on their behalf, provided adequate safeguards are in place. This is an issue which I cannot explore in detail here, however.

[133] Robertson (1991: 128).

[134] For instance, where an individual suffering from Alzheimer's disease appears to be debating different options, but on further enquiry, cannot recall why she is making the decision in the first place.

[135] Culver and Gert clearly express the rationale of this: "the primary point of determining competence is not to prevent others overruling patients whose decisions should not be overruled but to allow them to overrule patients whose decisions should be overruled" (2004: 262).

[136] This problem of ambiguity arises because it is not certain in what sense Robertson is using the term 'competency': "Competent people get to decide for themselves… Persons falling on the incompetency side of the line still deserve respect as persons, but respect for them does not mean that we honor their preferences or other expressions of choice" (1991: 128). This statement could be read as referring both to FPCLC and to decisional competence. In a later stage of the discussion in relation to medical treatment, Robertson equates competence with capacity for choice (*ibid.*: 140), but this does not establish that he is using competence in this way throughout.

preferences in relation to the decision legally recognised (such as the preference of a fifteen year-old to vote for a specific political party who has the decisional competence to reach such a decision). However, the fact that one is decisionally competent to choose but legally incompetent to make that decision for oneself does not mean that we must believe that the grounds for not ascribing FPCLC are stronger than the grounds for ascribing it.

Individual Autonomy

At the heart of any judgment of decisional competence or FPCLC is a normative conception of individual autonomy.[137] This normative understanding of autonomy has analogues in other areas where a legal recognition of decisional competences is made, such as freedom of conscience or freedom of contract. In order for consent not to collapse into coercion, consent must involve the absence of external control and a minimal level of autonomy consistent with its permissive function. An analytical concept of autonomy as self-determination thus gives consent its basic conceptual meaning and logical coherence.[138] A normative concept of autonomy enhances this analytical definition, insofar as it explains why consent should be valued apart from its logical coherence. A normative concept of autonomy thus emphasises the moral significance of consent.

At its simplest, the moral significance of consent amounts to a belief that those individuals who are at the centre of a decision that affects either only them or another consenting party get to make their own choices as far as possible, rather than have the choices of others imposed upon them.[139] The notion of 'harm to others' originally formulated by John Stuart Mill, typically circumscribes the realm of this choice.[140] In the *Groundwork of the Metaphysic of Morals,*[141] Kant argues for the value of individual autonomy as proceeding from the will of a "rational being". For Kant, autonomy is a necessary criterion for rational action, but simply because most human beings possess the capacity for rational action does not presuppose that they will employ it in all circumstances of deliberation. Various impulses and desires

---

[137] In Chapter 3, I will set out the normative concept of autonomy which I will apply in this book.

[138] Such as that embodied within Faden et al. sense$_1$ definition of consent.

[139] Faden et al. (1986: 288) concede this even within their analytical definition of consent: "[G]atekeeping by allowing autonomous persons - competent persons – to give informed consent and not allowing non-autonomous persons - incompetent persons – to give informed consent is accomplished by an appeal to the principle that autonomous persons are rightfully the decision-makers."

[140] Mill (1991).

[141] Kant (1991).

exist that blind the will from what is rational. These impulses lie outside the rational will. Acting upon them is to act without autonomy, or heteronomously[142]. Autonomous action entails accession to the laws one has made for oneself that can be subject to universalization:

> Autonomy of the will is the property the will has of being a law to itself... Hence the principle of autonomy is 'Never to choose except in such a way that in the same volition of the maxims of your choice are also present as universal law'.[143]

To be autonomous is, for Kant, simply to adhere to the requirements of the rational will. As such, "[a]utonomy is, therefore, "the ground of the dignity of human nature and of every rational nature".[144]

Gerald Dworkin reconceptualizes autonomy along different lines from Kant.[145] According to Dworkin, autonomy is "a second-order capacity of persons to reflect critically upon first order preferences..., and the capacity to accept or attempt to change these in the light of higher-order preferences and values".[146] It is in exercising autonomy, "persons define their nature, give meaning and coherence to their lives, and take responsibility for the kind of person they are".[147]

Both definitions of autonomy invoke values of universalizability, rationality, meaning and coherence that also inform the ethical discourse of consent. The concept of autonomy underpinning consent is universalizable in a Kantian sense, in that everyone who is capable of consenting should be morally and legally permitted to do so. However, consent does not depend upon avoidance of heteronomous motives for it to be valid. For instance, take an individual who seeks and consents to cosmetic plastic surgery. The motive is almost certainly vanity and therefore, according to Kant, an irrational desire or an impulse. Nonetheless, the expression of such a motive through consent will not make a difference to the validity of the consent given, at least in legal terms (and on most moral accounts). Equally, the consent would still be legally valid if the person consented to the procedure but

[142] Kant (1991: 102).

[143] Kant (1991: 101).

[144] Kant (1991: 97).

[145] Dworkin (1988). This is similar to the theory developed some years earlier by Harry G. Frankfurt (1971), who analyses the significance of "the capacity for reflective self-evaluation", exhibited in the ability to form "second-order desires" (1971: 7).

[146] (1988: 20). Dworkin illustrates this with the example of smoking (1988:15; 1989: 59). An individual may intend that she smokes a cigarette, and values the satisfaction that it brings. This is an example of a "first order preference". However, that individual may concurrently be aware of the substantial health risks incurred by smoking and desire that she gives it up. Alternatively, that individual may be aware of those health risks, decide that the satisfaction outweighs the risks, and continue smoking. These are examples of "second-order reflections". In this way, an individual can be autonomous according to Dworkin irrespective of whether her first-order preferences are compatible with her second-order reflections (1988: 15).

[147] (1988: 20).

wished secretly that she did not feel that her appearance was so important to her that she is prepared to devote a considerable amount of money and time to the surgery. Thus, the type of autonomy embodied in contemporary applications of consent is more Dworkinian than Kantian, insofar as the values, preferences and desires that motivate consent need not be strictly rational in a Kantian sense and may even be in conflict within a person.

The revival of normative political and social philosophy in the last four decades of the twentieth century has stimulated discussions of autonomy in bioethics.[148] Such discussions often taken place by juxtaposing autonomy with the practice of medical paternalism.[149] As a result, consent is presented as a means to respect and preserve decisional competence and has come to be associated with liberal values such as equality and rights. To suggest that consent and autonomy have become intertwined would not overstate the evolved trend in bioethics and biolaw, but nor should it imply that this trend has been seen as unproblematic,[150] nor diminish the equal importance of autonomous refusal of treatment or research participation.

Wear, in particular, believes that the importance of consent and autonomy have been exaggerated and are counter-productive to the goals of medicine:

> [D]o we really want to say that the notion of autonomy is so sacrosanct that even questioning it is offensive, and that the physician who feels that a patient is making a tragic, foolish, stupid, or self-destructive decision at most can offer to restate the case, rather than investigate whether there are actual and substantial flaws in the patient's decision-making process itself? To say this would seem to elevate the notion of autonomy to that of a fetish.[151]

Wear's appeal to regain a sense of perspective is a welcome antidote to the risk of "over-valuing" consent.[152] An acknowledgement of the value of consent should not marginalize the judgment of a health care professional in re-assessing decisional competence in respect of that matter. Indeed, it may serve to strengthen it. Apparent substantial flaws in an individual's decisional processes provide *prima facie* grounds for bringing her decisional competence to consent into question (particularly if she is a member of a group for whom FPCLC is presumed). They also provide grounds to err on the side of making a judgment of incompetence to decide (particularly if the health care professional is dealing with an individual whose FPCLC is not presumed in law but determined on a case-by-case basis, as is often the case with adolescents).

In the context of medical treatment, Wear goes on to claim that many patients who trigger a reassessment of their decisional competence will have done so on the

---

[148] Jonsen (1998, Chapter 3).

[149] Corrigan (2003: 769). An example of such a discussion can be found in Childress (1995).

[150] For critical discussions, see Bloche (1998), O'Neill (2002) and Corrigan (2003).

[151] Wear (1998: 136).

[152] Brownsword (2004), Beyleveld and Brownsword (2007a).

basis of behaviour or reasoning displayed in advance of the informed consent process.[153] It is true that the grounds on which assumptions of decisional competence may be brought into doubt include circumstances where the standard of decision-making required for the current decision-making context have been inadequately demonstrated in similar contexts. However, there is a risk of seeking to extrapolate too much from one decision-making context to another in order to justify a judgment of decisional competence in respect of the latter. There would have to be sufficient similarity between the two instances of decision-making at issue and the kind of decisional competences they require for extrapolations of this nature to be justified.

In the same way that a definition of decisional competence requires that an individual be informable to make that decision, a definition of decisional competence to consent requires that an individual receive sufficient information in order to make that decision in an appropriate way. For consent to be the valid expression of a reasoned choice, it must be offered with as much knowledge as possible of the act or procedure to which the individual is consenting. It is incoherent to suggest that voluntariness does not presuppose sufficient disclosure of this information. The absence of sufficient information disclosure ethically compromises any resulting consent, as it has been obtained through concealment and not because of an informed, reasoned choice.

The relationship between the capability to be informed and information provision can often lead to the latter establishing the presence of or raising questions about the former. There are good grounds to doubt decisional competence, if in the process of seeking consent, a physician or researcher observes that the potential participant appears incapable of retaining, understanding and evaluating the information supplied to her when presented in a suitable manner. The provision of sufficient information about the decision is therefore particularly useful for observing instances of decisional competence where there has been no previous contact between the person seeking consent and the person from whom consent is sought.[154] In this way, the consent processes builds in safeguards that serve to monitor decisional competence. If the informed consent process is undertaken thoroughly and sincerely, the chance of individuals being presumed competent who are actually decisionally incompetent (a so-called 'false-positive' judgment of competence[155]) and vice-versa is much reduced. Avoiding false-positive judgments of decisional competence is especially important with groups whose decisional competence is tenuous or questionable.

---

[153] Wear (1998: 142)

[154] Such as in the context of biomedical research, where the first contact between a clinician or researcher and potential participant may well be the informed consent process.

[155] "The more the downside risks associated with a false positive determination of competence increase, the more one will wish to make sure that a patient is in fact choosing competently" Cutter (1991: x–xi).

All decisions for which consent is required determine their own level of what constitutes sufficient information, but in any case, it must at least meet a reasonable standard of completeness and intelligibility. This is a particular challenge where the relevant information is complex or extensive. By way of illustration, consider the intrinsic difference between therapy and research. The nature of biomedical research typically involves intentions, procedures and outcomes that are more complex than those of most forms of treatment.[156] This gives rise to a *prima facie* recognition that criteria for valid consent to research will usually require a higher level of decisional competence than for treatment. Without this recognition, decisional competence to consent to research becomes indistinguishable from decisional competence to consent to treatment, negating the fact that the purpose of research raises medical and ethical implications for the potential participant that are wholly distinct from therapy.

The two most significant characteristics of research are the potential harm that may result from participation and the principal motive – the advancement of scientific knowledge, irrespective of whether the research has the possibility or aim of directly benefiting the participant or how much the interests of the participant are to take precedence over scientific endeavour.[157] These implications increase the significance attached to 'false positive' presumptions or judgments of decisional competence that are used as the basis for an ascription of FPCLC. To avoid this, the informed consent process needs to distinguish clearly the aims of therapy and research in such a way that, in consenting to the research, the potential participant does not mistake research for therapy.[158] If she appears unable to comprehend this distinction, then there are compelling grounds for reaching a judgment that she is incompetent to make a decision to participate on her own behalf.

Why Biomedical 'Research'?

Reference to biomedical 'research' is often taken to be unambiguous. The use of this term stands in need of justification, however, particularly when juxtaposed with the more evocative terminology of 'experimentation'.

In general terms, experimentation describes a procedure or investigation designed to test a hypothesis where it is not possible to predict the results. Experimentation has been used to describe methodologies as diverse as those in

---

[156]Examples of information that potential volunteers may be required to understand in order to have decisional competence to consent to biomedical research include specialised or philosophical concepts, such as randomisation, altruism or the function of a placebo. See Foster (2001: 57).

[157]To suggest otherwise would be to deny the difference between research and innovative therapy, whose principal motive is patient benefit.

[158]Known otherwise as the 'therapeutic misconception'. See Appelbaum et al. (1982).

contemporary music composition[159] and to empirical lines of inquiry in the social sciences[160] – it is not merely confined to the medical sphere. One may take research and experimentation to be synonymous, although in its vernacular sense, experimentation often connotes how a line of enquiry is addressed rather than the nature of the outcome it generates.

Attempts at definitions display a piecemeal lineage. At the beginning of the nineteenth century, Thomas Percival claimed that medical experimentation takes place where existing medical practices prove unsuccessful and under circumstances not previously foreseen.[161] Percival's observation identifies the motive for overcoming the limits of existing knowledge and reinforces the unknowable quality of the outcome, although it does not speak of anticipated outcomes that may attach to the procedure of research design.

Almost 150 years later, McCance endorsed a definition of medical experimentation as a procedure undertaken with a subject which is not accepted by medical practitioners as offering primary therapeutic benefit or assisting the process of diagnosis and where the results cannot be known in advance.[162] This essentially reframes Percival's definition and offers little new insight. At around the same time, the World Medical Association stipulated in the *Declaration of Helsinki* what types of medical research involving human participants there are, without stating what medical research actually is.[163] The guidance makes explicit reference both to experimentation and to research whilst using the terms synonymously.

A decade later, Morgenbesser attempted a more streamlined definition, still couched in the language of 'experimentation':

> an undertaking is called a biomedical experiment…if it is instituted to gather data in a statistically significant way which may be used to test a medical hypothesis, or more generally, to test a medically related hypothesis… if it is of the form 'If S then…' where 'S' stands in place of an action which doctors can be expected to be able to undertake and execute[164]

More recently, McNeill expressed a preference for experimentation to research, in order that the definition include forms of innovative treatment:

---

[159] Most notably, John Cage has employed 'aleatory' or 'indeterminate' techniques such as *I Ching*, which rely wholly or partly on chance (Cage, 1961). Such chance-based approaches would be inappropriate for research with humans, however, as they treat the experimental subject material in purely instrumental terms and do not possess a sufficiently clear *a priori* methodology to meet with the approval of ethical review.

[160] Such as 'covert observation', a research methodology in which data is collected from unwitting participants, oblivious to the fact that research is being carried out. The importance attached to consent in research ethics has now discredited such methodologies.

[161] Percival (1803).

[162] McCance (1951), reproduced in Reiser et al. (1977: 275).

[163] World Medical Association (1964, rev. 2000) *Declaration of Helsinki: Ethical Principles for Research Involving Human Subjects*, Part A 1.: "Medical research involving human subjects includes research on identifiable human material or identifiable data".

[164] Morgenbesser (1977: 100–101).

> The distinguishing feature of experimentation is that something new is being tried. Experiments necessarily carry with them unknown consequences. Once the consequences are known, they are known, they are no longer experiments.[165]

The issue here does not lie with how we construe human experimentation. In descriptive terms, this is relatively unproblematic – the above definitions are complementary rather than in conflict. The issue instead is whether it is important whether the appellation biomedical *research* or biomedical *experimentation* is used. Neither Percival nor McCance provide reasons as to why any procedure that is not designed to alleviate a condition should be considered experimentation rather than research, given that experimental medicine, like research, may in part have therapeutic benefits. Similarly, neither Percival nor Morgenbesser offer any grounds why experimentation could not be substituted with biomedical research when discussing the testing of a medical hypothesis. McNeill does not address why, in preferring experimentation to research, a definition of research cannot encompass research with a therapeutic dimension that extends beyond research trials, and what it is about using 'experimentation' rather than 'research' that would emphasize the importance of participants' protection by ethical review bodies.[166]

Of all the above definitions, McNeill is the only commentator who explicitly rejects 'research' in preference to 'experimentation'. His rejection of the term 'research' appears more a measure of his dissatisfaction of the use of the term by Levine,[167] who according to McNeill, considers that *most* medical practice is experimental. In seeking to distinguish his definition from that of Levine, McNeill appears to have taken a view of research which is needlessly restrictive.

Why, then, use biomedical 'research' if 'experimentation' will suffice? The use of biomedical research serves to clarify and distinguish the subject of inquiry. First, the language of research has come to predominate references to experimental procedures on human participants. The Declaration of Helsinki,[168] the International Ethical Guidelines for Biomedical Research Involving Human Subjects[169] and the Council of Europe Additional Protocol to the Convention on Human Rights and Biomedicine on Biomedical Research,[170] amongst others, bear witness to the frequency with which biomedical *research* is used in place of biomedical experimentation. For the sake of consistency alone, the term 'biomedical research' is preferable to biomedical experimentation. Second, the language of experimentation with human participants has become tarnished by its association with ethically barren medical 'experiments' conducted during the twentieth century. Research is thus preferable on symbolic grounds.

[165] McNeill (1993: 12).

[166] McNeill (1993: 12).

[167] Levine (1988).

[168] World Medical Association (1964, rev. 2000).

[169] Council for International Organizations of Medical Sciences (2002).

[170] Council of Europe (2005).

In this book, the descriptive element of biomedical research is taken to refer to all types of clinical investigations that have as their ultimate aim the pursuit of clinical knowledge, including those that have a partial therapeutic intent and those that do not. This includes innovative treatments, neuroscientific and psychiatric studies, randomised clinical trials (RCTs), and research involving new genetic technologies.[171] The normative element of biomedical research is taken to denote investigations that would meet and surpass the standard required of ethical review in a legal and political system founded upon human rights,[172] unless specified otherwise.

The Evolution of the Rights of Research Participants

The rights of volunteers and potential participants in biomedical research have evolved through three distinct stages. Each stage has overlapped or intersected with the next, but the evolution of a new understanding of rights in the research context distinguished each. In assessing the current status of the rights of research participants, more light can be shed upon the extent to which ascriptions of FPCLC are an integral part of contemporary guidelines.

The traditional conception of a *negative* right not to be entered into to biomedical research unless express consent was given constituted the first stage. This premise of this idea is the normative priority of informed consent as the central principle of biomedical research ethics.[173] Negative rights against participation animated the debate from the late 1940s, at least up until the 1970s and still have influence now. Seminal examples include the Nuremberg Code (1947) and the Declaration of Helsinki (first approved in 1964).

The second was a stage, still ongoing in many respects, where rights in medical research were framed predominately in terms of the *duties* owed to a participant by the researcher once she had validly consented to participate. These include, for example, a high standard of information disclosure, the right to withdraw from the research at any time and proper arrangements for maintaining the confidentiality of the results pertaining to patients and volunteers.[174] These duties attempt to reconcile

[171] It is not possible to discuss all the contemporary forms of biomedical research that exist, but see n. 18 above for an overview of those relevant to the focus of this book.

[172] One that would be compliant with the requirements of the PGC. See Chapter 3 and Beyleveld and Brownsword (1986: 304–314, 392–393).

[173] Fletcher (1983: 210) has questioned whether consent should be the primary animating principle of research ethics: "the evolution of post-Nuremberg ethics of research began with a highly symbolic and rigid display of reasoning about consent. This was perhaps appropriate for showing moral disgust, but hardly sufficient for the *first principle*." This objection overlooks the semantic importance of consent in distinguishing the moral permissibility of biomedical research with human participants from morally impermissible forms of scientific investigation. See Chapter 3 for discussion.

[174] See Fulford and Howse (1993).

the imperative of protecting human participants with the value of efficiency in the conducting of medical research.

The third is the emergence of a *positive* right to involvement in biomedical research whereby the choice whether or not to participate is treated as a social right exercisable by the rights-holder.[175] Such thinking is epitomised in the preamble to the Additional Protocol to the Convention on Human Rights and Biomedicine on Biomedical Research,[176] and in the ethical guidance issued by Alzheimer Europe, who have stated, "people with dementia have a right to participate in research, should they so desire".[177] Others have gone further, suggesting that more inclusive research strategies are important to "gather the experiences and views of people with dementia themselves, rather than (or in addition to) those of proxies", which is important to "challenge current inequalities in social relations for people with dementia".[178] This position is also endorsed in the General Medical Council's 2002 guidance, Research: The Role and Responsibilities of Doctors, which suggests that the exclusion of vulnerable groups from medical research "could be a form of discrimination".[179]

The evolution of the rights of research participants has shifted the emphasis in biomedical research with human participants from conventional protectionism towards empowerment and in doing so has recast protectionism *as* empowerment. By giving individuals an active role in the selection of human research participants, the power imbalance between researcher and participant narrows. This may give rise to a 'paradigm shift' in which those most at risk of exploitation in biomedical research are accorded enforceable rights of choice, mirroring the earlier transition in values underpinning medical treatment from paternalism to autonomy.

The more protection that individuals have to decide for themselves – particularly those whose decisional competence may appear to be questionable – the more difficult it is for others to take decisions on their behalf about whether or not they should participate. We are then confronted with the related question of whether biomedical research ethics *ought* to be moving in the direction of encouraging positive rights, at which point the limits of an analytical understanding of the rights of research participants is reached. Before we begin considering the normative grounds on which individuals with questionable decisional competence to consent ought to be permitted to consent to or refuse research participation, an understanding of the nature of vulnerability and its relevance in biomedical research is important.

---

[175] Neves (2007) also identifies such a stage, which she calls the 'vindicatory stage' (*ibid.*: 184).

[176] Council of Europe (2005).

[177] Alzheimer Europe (undated).

[178] Wilkinson (2002: 12–13).

[179] General Medical Council (2002, para. 43).

The Meaning of Vulnerability

How we think about vulnerability and the conditions that give rise to it have a direct impact upon how a legal system may in turn define the status, govern the relationships and guarantee the needs that are associated with being vulnerable. The perception of our own vulnerability and that of others within a folk-psychological context influences our thinking in this regard. That vulnerability has been until recently a notably under-theorized concept in bioethics, as well as in moral, political and legal philosophy, may be somewhat surprising, given its significance.[180] We should not take this to mean that vulnerability is of no interest to bioethicists or moral, political and legal philosophers,[181] but that the meaning of vulnerability is considered less central and open to fewer interpretations than other bioethical, moral, political and legal concepts, such as consent, personhood, justice or rights. This inattentiveness overlooks the normative importance of vulnerability and its permutations, which are relevant to bioethics in general and to biomedical research ethics in particular.[182]

The word, "vulnerable" has its origins in the Latin verb *vulnerare*, to wound. This original meaning is still reflected today, as the *Oxford English Dictionary* still lists "susceptible of receiving wounds or physical injury" as a primary definition of vulnerability.[183] In order to think conceptually about vulnerability, we need to explore the semantic resonance that extends beyond propensity to physical harm. Vulnerability also includes a predisposition to certain types of psychological and/or developmental harm that an individual has an interest in avoiding. It is more accurate to conceive of vulnerability as denoting a holistic set of fundamental interests, which, if not met through neglect or abuse, give rise to harm, deprivation or suffering. These interests include at the most fundamental level, life, physical health and mental equilibrium necessary for agency at all.[184] They extend to encompass other universal interests such as food, shelter, education, healthcare and bodily integrity.[185] On this account, it is possible to conceive of vulnerability as connoting the

---

[180]MacIntyre (1999, Chapter 1) and Levine et al. (2004) share this view. It is further illustrated by the fact that in the current edition of the authoritative two-volume reference work, *Encyclopaedia of Ethics* (Becker and Becker, 2001) and in the recently published second edition of *Principles of Health Care Ethics* (Ashcroft et al., 2007) 'vulnerability' is not included as an entry nor as a chapter topic respectively.

[181]A notable recent exception is Hoffmaster (2006), who suggests that moral philosophy has marginalized vulnerability due to its antagonism with autonomy and individualism. MacIntyre also pursues this theme in his Carus lectures, published as *Dependent Rational Animals* (1999). Elsewhere, Silberfeld (2002) considers vulnerability but as a quality of incompetent persons only. O'Neill (2000) and Anderson and Honneth (2005) invoke vulnerability but do not subject the concept to any sustained form of conceptual analysis. Goodin (1985) approaches vulnerability as a determinant of welfare need, rather than an analysis of the concept in its own right.

[182]Apart from Hoffmaster (2006) and Silberfeld (2002), Hurst (2008), Zoboli (2007), Neves (2007), Kipnis (2003, 2004), Kottow (2003, 2004) and Weisstub and Thomasma (2002) have recognised the importance of theorizing vulnerability for bioethics.

[183]Oxford English Dictionary, second edition (1989).

[184]Gewirth (1978:54).

[185]Gewirth (1978:54).

perennial threat to these shared fundamental interests all human beings face *in the same way*, which, if or when such threats materialise, gives rise to harm or suffering. Although vulnerability does not always connote harm that is inexorable or certain, it is impossible to imagine a life that is not in some way affected by the threat posed to such interests.

Human frailty, fallibility and mortality need to be accounted for when conceptualising vulnerability. The very fact that one's capacity for agency can be easily restricted or ended altogether by others and our environment is a universal experience of the human condition and renders us all, in the same way, vulnerable.[186] Vulnerability, therefore "... expresses the condition of all life as able to be hurt, wounded and killed".[187] The recognition of this vulnerability is manifest by the 'existential anxiety' that surrounds the possibility of our own extinction.[188] Insofar as this quality applies to all human beings, it is a quality that one all too often forgets applies to oneself when identifying 'the vulnerable' – after all, few people choose to be conscious of – let alone celebrate – their own vulnerabilities.

When defining vulnerability, two levels of meaning emerge. The first is the common experience of "existential vulnerability"[189], which I call 'baseline vulnerability'. The second relates to those individuals who for some reason experience a *heightened* state of vulnerability. On this account, vulnerability is not conceived of on an 'all or nothing' basis, but rather on a 'more than usual' basis.[190] This nuances the concept of vulnerability in such a way to avoid Levine et al.'s criticism that "[i]f everyone is vulnerable, then the concept becomes too nebulous to be meaningful".[191] Indeed, it is precisely because everyone is vulnerable that we need to be able to identify cases where it is especially pronounced.

Conditions of mutual unconcern aggravate heightened vulnerability. Undeserved differences in circumstances and abilities along with the vagaries of chance contribute to inequality and exploitation when societal and legal institutions do not address them. This applies also to baseline vulnerability. It is quite apparent how neglect of baseline vulnerability can prove sufficiently deleterious that instances of baseline vulnerability over time become instances of heightened vulnerability. The principal

---

[186]Beyleveld (2001).

[187]Kemp (1998: 5).

[188]Beyleveld and Brownsword (2001: 117). In this context, one is reminded of the poignancy of Roberto Unger's aphorism, "Salvation through the acceptance of vulnerability is the only kind of salvation there really is" (1984: 164).

[189]Thomasma and Pellegrino (1998: xxi) also use the term 'existential vulnerability', but do so in a different way to refer to a "disease state" (ibid.).

[190]George W. Harris (1997) argues that it is in our vulnerability to character breakdown that the nature of human dignity is manifest. Harris uses the example of a mother forced to choose which of her children should die so that the other may live. He claims that for this woman to be vulnerable to anguish (and eventually suicide) is an expression of her dignity. However, this seems to suggest that in order to act with dignity in the face of vulnerability we have only one inexorable course of action available to us, which may lead to the end of capacity for agency altogether. This is a thesis which sits uneasily with the conception of human dignity arising out of the free will of agents and the resultant possibility of responding to psychological suffering in a way which does not consume us.

[191]Levine et al. (2004: 46).

aim of law in relation to vulnerability within a socially just society is, therefore, twofold. It must be both ameliorative (i.e. to provide for outcomes that mitigate the effects of baseline and heightened vulnerability, through healthcare, education, welfare, social inclusion, etc) and protectionist (i.e. to safeguard the important interests of those who for whatever reason experience heightened vulnerability that cannot be mitigated through ameliorative measures).

We can explain heightened vulnerability by reference to a range of cognitive and circumstantial factors.[192] These include immaturity, old age, physical illness or injury, mental illness or impairment, lack of education, socio-economic disadvantage (including both lack of resources and opportunities), trauma (physical or psychological), institutionalisation (punitive and therapeutic) or membership of a minority group that experiences prejudice or persecution. The possession of one or more of the above characteristics typically predisposes the individual to heightened vulnerability through dependence, disempowerment or other social inequality associated with that condition or with those circumstances. The impact that heightened vulnerabilities have upon the ability of the person to act successfully and independently in the world around them emphasises this and the broader *relational* dimension of vulnerability.[193] At the extreme, a failure to be able to engage at a basic level with the immediate environment gives rise to a sufficient range of vulnerabilities to 'lock out' the individual from a world of independent interaction.[194]

As the two forms heightened vulnerability may take are not mutually exclusive and may overlap,[195] it is helpful to identify in which respect individuals are primarily vulnerable, particularly for the purposes of identifying vulnerability in research participants. Individuals with mental disorder or intellectual disabilities and children are primarily *cognitively vulnerable* (albeit to different degrees) due to incompleteness or imbalances of their perception, ratiocination and affect that can limit or undermine task and decisional competences. These are intrinsic vulnerabilities insofar as their source lies within the agent's being. The economically disadvantaged, prisoners, the uneducated and persecuted are primarily *circumstantially vulnerable* as the circumstances that make them vulnerable are contingent upon human activities, especially social, political and legal arrangements. These are extrinsic vulnerabilities insofar as their source lies outside of the agent's being.[196]

---

192 Berg et al. (2001: 266).

193 See Thomasma and Pellegrino (1998:xx–xxi) for a discussion of relational vulnerability in the context of health care.

194 In many ways similar to societal incompetence.

195 Berg et al. (2001: 266), Kipnis (2004: 230).

196 It should be stressed that this conceptualisation of vulnerability does not seek to valorise the social milieu within which heightened vulnerabilities are understood. The challenge to the medical model of disability offered by the social model and by the Disability Rights Movement yields many valuable insights for bioethics. Whilst I do not consider the taxonomy of vulnerability I have presented in this context as 'medicalising' those vulnerabilities, I am aware it could be open to dispute in others, which would require a lengthier discussion than is possible here to properly address. For an illuminating Foucauldian critique of 'disability' and 'madness' that identifies these contexts, see Tremain (2002).

Cognitive vulnerability may denote a permanent or temporary quality of the individual's mental state. It is enough to recognise at the particular point in time we investigate an individual's decisional competence that she either currently is or has been affected at some stage by a mental disorder that could lead to cognitive vulnerability.[197] However, we cannot conceive of cognitive vulnerability purely in terms of a fixed, abstract concept. Rather, it is a protean and manifold experience that requires detailed explanation and analysis of its many forms, albeit according to normative criteria, such as that which supplies the basis of psychiatric diagnosis.[198] Of course, it is possible to accept that circumstantial vulnerabilities could themselves exacerbate or even create cognitive vulnerabilities (such as in the case of a terrorist suspect subject to indefinite detention who experiences severe psychological suffering as a result), to the extent where the line between circumstantial and cognitive vulnerabilities blur. Nevertheless, because we can still identify the origin of the heightened vulnerability, the distinction between cognitive and circumstantial vulnerability remains a useful heuristic device.

There is persistence in many experiences of vulnerability, which, especially in cases of cognitive vulnerability, is difficult to address. O'Neill argues that through an acceptance of the vulnerability of all, it may be possible to contain vulnerability where it is most prominent.[199] This argument endorses an institutional response towards vulnerability that is an ideal-typical feature of social democracy. However, it does not sufficiently account for what to do with cognitive vulnerabilities such as mental illness or intellectual disabilities, which legal or political responses alone cannot ameliorate.[200] The challenge to accept such persistent vulnerabilities and to consider the institutional responses that are appropriate to them then confronts us.

We need to be clear about what kind of values should guide these responses. Broadly speaking, we could choose to apply one of three types of value framework.

---

[197]This will probably, although not necessarily, have been diagnosed with reference to one of the standard nosological systems, such as DSM-IV or ICD-10.

[198]See Sadler (2005) for a discussion of the relationship between value judgments and psychiatric diagnosis.

[199]O'Neill (2002: 88–89).

[200]Michael Kottow (2003, 2004) draws a distinction between vulnerability and susceptibility in order to emphasise this point: "[v]ulnerability can be reduced by equal protection to all members of society under a principle of justice. Susceptibility is a determined state of destitution and can therefore only be reduced or neutralised by measures that are a) specifically designed against the destitution in question and b) actively applied" (2003: 463). Kottow conceives of vulnerability as something which can be reduced by (presumably) state or other collective action. For vulnerabilities to be ameliorated in this way and be of benefit to all suggests that Kottow believes vulnerability can be ameliorated through something akin to a principle of equality before the law. The equation of susceptibility with destitution is to define susceptibility as being remedied through welfare rights. In doing so, there is no apparent scope for cognitive vulnerability. In drawing this distinction, Kottow has not produced a framework analogous to cognitive and circumstantial vulnerability, but instead to baseline vulnerability and circumstantial vulnerability. This framework is less facilitative of thinking about different types of vulnerabilities than the one advanced here.

The first type, autonomy-centred frameworks, tends to endorse ascriptions of decisional and task competence as far as possible on behalf of vulnerable individuals. The second type, paternalism-centred frameworks, tends to endorse ascriptions of decisional and task incompetence as far as possible on behalf of vulnerable individuals. The third type, which represent a hybrid of the above, advocate paternalistic behaviour only as an aid to the expression of decisional or task competences or to protect those whose decisional competence is irreparably hindered. Whichever framework we favour, we must be consistent in applying it. To apply a different value framework to the same cognitively vulnerable group in similar circumstances without adequate justification risks giving rise to incoherence and illegitimacy.[201]

Deciding whether an institutional response to cognitive vulnerability should be guided purely by autonomy, paternalism or both should turn upon whether: (a) the individual or groups of individuals appear to be able to exercise agency, and if so; (b) can benefit from the empowerment proposed. Consideration of (b) is only possible in the presence of (a).[202] Judgments of benefit require empirical evidence as well as moral argument. This is in order to avoid a situation where moral arguments justify treating individuals with persistent cognitive vulnerability as if they have baseline vulnerability by presuming the same potential for decisional and task competences and holding them as fully to account for their choices, when available empirical evidence suggests that they do not have and cannot develop these competences. Were such a situation to arise in the context of medical research, we would not be protecting the autonomy of the individual through empowerment but instead expecting the individual to make a choice that she is probably mentally incapable to make, irrespective of how much information the researcher provides and how sensitively the researcher seeks consent. This is analogous to expecting an asthmatic non-swimmer to swim 100 metres after the principles of swimming had been explained to her simply because she appeared in all outward respects physically fit.

Cognitive Vulnerability and Consent to Biomedical Research

Biomedical research of any kind will pose a certain degree of risk to all participants, irrespective of whether they experience baseline or heightened vulnerability. However, many potential participants in biomedical research who are cognitively vulnerable will have questionable decisional competence to consent, and as such, may neither have a rebuttable presumption of FPCLC enshrined in law nor present clear grounds for a judgment of decisional incompetence. Their participation in

---

[201] Consider, for example, the current position in England and Wales where it is legally possible at sixteen years of age to join the armed forces and have a sexual relationship but not to elect the Government and have the final say in refusing medical treatment.

[202] Because it does not make sense to say that non-agents can be empowered.

research thus gives rise to additional ethical and legal dilemmas that typically are not associated with adult potential participants in good mental health or of average intelligence.[203] By the same token, the possibility of judging decisional competence is not a realistic one with potential participants who are manifestly unable to consent, such as neonates, infants and young children, the comatose and individuals in a persistent vegetative state. The ethical questions that exist in relation to their participation surround proxy consent or surrogate decision-making.[204] Such questions lie outside the scope of this book.[205]

The participation of prisoners and socio-economically deprived individuals (both in relative and absolute terms) in biomedical research also raises separate questions about whether circumstantial vulnerability undermines their decisional competence to offer informed consent. Incarceration, institutionalisation and poverty are factors external to the cognitive function of the individual that increase vulnerability and potentially undermine decisional competence, though these factors may still bear upon cognitive function. Of course, it will be the case that some prisoners and socio-economically deprived individuals will also experience mental disorder or intellectual disability or will still be in adolescence. Here, their primary source of heightened vulnerability is cognitive, even though they also experience circumstantial vulnerability. Such individuals fall within the scope of potential research participants under consideration in this book. Were the sources of circumstantial vulnerability removed, they would still be very likely predisposed to experience cognitive vulnerability. This is because the sources of cognitive vulnerability are much more resistant to removal due to their embeddedness within the psyche of the individual. In these cases, we can more accurately understand the sources of circumstantial vulnerability as exacerbating the individual's overall level of heightened vulnerability rather than being exclusively determinative of it.

When we encounter questionable decisional competence in research contexts, the higher threshold of understanding required compared to that for therapeutic contexts, fused with the risk of harm to self, often resolves the question of decisional competence in favour of incompetence. Once incompetence is presumed, any decision to participate becomes a matter for a surrogate decision-maker and is commonly resolved by recourse to considerations of 'benefit' or 'negligible/minimal harm', stipulated in ethics guidance or in law, usually determined by a research ethics committee and applied by a research investigator and/or the incompetent individual's surrogate decision-maker. This approach may indeed be justified where there is satisfactory evidence to reach a judgment of decisional incompetence to

---

[203] This does not mean that problems never arise, but they tend not to reflect on the groups as a whole. Where adults presumed decisionally competent to consent by an ascription of FPCLC are found on subsequent examination to lack the requisite decision-making abilities to consent, this is typically explained by subsumption within the categories of mental disorder or intellectual impairment – two central categories of cognitive vulnerability.

[204] Or, in the case of decisionally incompetent adults, advance directives for research. See Berghmans (1998).

[205] See Buchanan and Brock (1990) for a thorough treatment.

consent, but is less justified as an appropriate first response for instances where available evidence suggests that the individual in question possesses decisional competence, although is not presumed to have FPCLC to make this decision.

Such instances reveal a tension that relates to uncertainty about whether a presumption in favour of decisional competence or decisional incompetence ought to be applied in relation to specific groups of cognitively vulnerable individuals, or whether instead a case-by-case approach should be adopted in making such judgments. This can only be resolved in accordance with a clear normative framework. However, in doing so, we should not lose sight of the particular risks which participation in biomedical research present to cognitively vulnerable individuals. The requirement to satisfy the five general constitutive abilities of decisional competence in relation to the type of research proposed will be demanding even for an adult with a typical range of decisional competences. For an individual with cognitive vulnerability, this will be especially challenging. Impairment of or underdevelopment in understanding, reasoning, deliberation, communication or capacity to form voluntary purposes that are of value will undermine the integrity of the decision-making process and will increase the likelihood of the cognitively vulnerable individual not choosing autonomously. This, in turn, gives rise to a risk of the cognitively vulnerable individual being judged competent to make a decision that involves some form of interference with her fundamental interests in bodily or psychological integrity, when this is not the case. Such erroneous 'false positive' judgments are especially problematic where the ethical ramifications of erring in favour of the presence of competence are great.

The importance of safeguarding these fundamental interests against encroachment warrants a meticulous degree of scrutiny to ensure that the judgment made about the decisional competence of the cognitively vulnerable individual is as complete and reliable as possible. Thus, in cases where we are to treat the cognitively vulnerable individual as decision-maker (rather than a surrogate), we need to be as certain as we can that she is able to engage effectively in the relevant decision-making process, particularly where one of the possible choices it offers (such as research participation) involves potential harm to these fundamental interests.[206] This requirement for a high degree of certainty will be justified by all normative frameworks that take human well-being seriously.

By contrast, the additional risks posed to circumstantially vulnerable individuals through participation in biomedical research turn upon to what extent the oppressive environment erodes opportunities for deliberation and the maintenance of voluntary purposes relevant to decision-making about participation. Coercion, or more

[206] As discussed in Chapter 1, the relevance of risk here is one that attaches to the entire decision-making scenario, irrespective of the choice reached (consistent with a decision-relative approach to decisional competence favoured in this book), rather than contingent upon the choice reached (consistent with a risk-relative approach to decisional competence).

subtle forms of pressure to participate, may overwhelm their ability to deliberate free from malign influences and to form purposes which are unaffected by a desire or need to satisfy the wishes of the dominant party. The existence of subordinate relationships between the circumstantially vulnerable individual and the individual(s) or institution responsible for conducting the research may tend to generate such pressures. In the context of psychiatric research, for example, supporters of what has become known as the 'psychiatric survivor' movement have questioned the very possibility of genuine informed consent to research participation within the confines of the psychiatric hospital.[207]

Whilst the theoretical basis upon which ascriptions of decisional competence turns is mental functioning rather than apparent heightened vulnerability or legal status, the presence of decisional competence may be less readily doubted where a potential participant's heightened vulnerability is circumstantial (such as being imprisoned) rather than cognitive (such as being a child). This is because of an understanding of the way in which the two different forms of heightened vulnerability serves to impede decisional competence. Whereas the development and exercise of decisional competence would usually be possible in the absence of the environmental conditions which generate circumstantial vulnerabilities (assuming no underlying or resultant cognitive vulnerability), it is more difficult for environmental modifications alone to assuage the impact of cognitive vulnerabilities on decisional competence, unless their purpose is specifically to elicit and support decision-making skills.

Discussions of vulnerability have received more attention in certain codes of biomedical research ethics than in the academic literature on bioethics in recent years, although some of the major codes conspicuously omit a definition or analysis of vulnerability.[208] Regrettably, some of those that have made explicit reference to vulnerability are underdeveloped in this regard. For example, the 1998 Barcelona Declaration: Basic Ethical Principles in Bioethics and Biolaw defines vulnerability narrowly in terms of a quality that all agents share (which correlates to baseline vulnerability), without indicating when and how vulnerability might be heightened. On this basis, there are no grounds for making judgments of relative vulnerability and tracing its origins.[209] The definition of vulnerability used in the slightly earlier *ICH Harmonised Tripartite Guideline for Good Clinical Practice* (issued in 1996) appears to omit decisionally competent cognitively vulnerable adult participants altogether, unless they are also experiencing some form of circumstantial vulnerability.[210]

---

[207]Morrison (2005: 73–74). Chamberlin (1977) offers a seminal account of the main aims and concerns of the movement.

[208]For example, two of the most significant codes, the Nuremberg Code and the Declaration of Helsinki are silent on what constitutes vulnerability in research populations. In English law, The Medicines for Human Use (Clinical Trials) Regulations 2004 utilizes the language of vulnerability without stipulating what 'vulnerable' means. See Schedule 3 Part 1 s.1 (i).

[209]Partners in the BIOMED-II Project: 'Basic Ethical Principles in Bioethics and Biolaw' (1998: 6).

[210]International Conference on Harmonisation (ICH) (1996: 8, para. 1.61).

The 2002 CIOMS International Ethical Guidelines for Biomedical Research Involving Human Subjects offers a more cogent definition, stating:

> vulnerability refers to a substantial incapacity to protect one's own interests owing to such impediments as lack of capability to give informed consent, lack of alternative means of obtaining medical care or other expensive necessities, or being a junior or subordinate member of a hierarchical group.[211]

This definition accommodates a more comprehensive range of vulnerabilities that are specifically relevant to the context of biomedical research. The corresponding guideline describes vulnerable persons *qua* persons as "relatively (or absolutely) incapable of protecting their own interests"[212]; *qua* potential participants in biomedical research as having "limited capacity or freedom to consent or decline to consent".[213] The CIOMS guidelines emphasise that a 'special justification' is needed in order to invite vulnerable populations to serve as participants in research and that means for protecting their rights and welfare must be applied.

The liberal approach of the 2002 CIOMS guidelines contrasts starkly with the protectionist approach taken toward the use of vulnerable participants in biomedical research contained in the 1979 US Belmont Report, which speaks of the 'injustice' that occurs as a result.[214] The examples of vulnerable participants given within the definition, namely "racial minorities, the economically disadvantaged, the very sick, and the institutionalised"[215] are, with few exceptions, groups of individuals with primarily circumstantial vulnerability. The Belmont Report does not mention adults with mental disorder or intellectual disabilities, although one might reasonably infer that, as the 'very sick' often experience cognitive vulnerability, these groups were also within the contemplation of the drafters of the Report. Thirty years on, the rigid prescriptions it advances appear anachronistic in light of how the rights of research participants have evolved.

The Explanatory Report to the Additional Protocol to the Convention on Human Rights and Biomedicine Concerning Biomedical Research, adopted by the Council of Europe in 2005, contains the most detailed taxonomy of vulnerability in a contemporary ethical code. It provides a rich taxonomy of vulnerability in research according to cognitive, situational, institutional, deferential, medical, economic, and social factors,[216] similar to Kipnis's taxonomy of vulnerability in research.[217] By its own admission, the Explanatory Report is not an authoritative interpretation

---

[211]CIOMS (2002: 10).

[212]CIOMS (2002) Guideline 12.

[213]CIOMS (2002) Guideline 12.

[214]National Commission for the Protection of Human Subjects of Biomedical and Behavioral Research (1979).

[215]National Commission for the Protection of Human Subjects of Biomedical and Behavioral Research (1979).

[216]Council of Europe (2004, para. 69).

[217]Kipnis (2004: 218, 222–230).

of the Additional Protocol.[218] However, it suggests the lines along which bioethical thinking about vulnerability is informing codes of biomedical ethics and goes some way to demonstrating that vulnerability is being taken seriously as a substantive concept in European bioethics and biolaw.

One area where cognitive and circumstantial vulnerabilities clearly overlap is in the participation of elderly persons in nursing or care homes in research. Reich argues that "elderly persons in nursing homes... are captive populations subject to exploitation" and limited in their ability to make free and voluntary decisions and "who should not be made victims of their own altruism".[219] For Reich, the considerable degree to which such individuals are vulnerable automatically divests them of decisional competence. Thomas and Waluchow[220] take a less severe attitude, but like White earlier, caution against the use of a legal definition of competence, in this context for selecting elderly participants for research:

> [O]ne should be wary of employing, consciously or otherwise, the legal model in selecting subjects for experimental research. It would be a mistake to infer that because an elderly person has not been shown to be fully incompetent she is therefore fully competent, and to conclude from this that one therefore need not worry at all about the capacity for informed consent. If there are many points lying between full competence and full incompetence, and if a person just might occupy different points at different times, then any such inferences would be invalid and fraught with danger. One would, in drawing such a conclusion be sweeping under the rug the hard choices that must be made, and simply ignoring potentially disastrous effects. There is, of course, great temptation to do this. It is far more difficult to establish degrees or grades of competency than it is to establish outright incompetence. It may be even more difficult to determine where along the spectrum lying between the two extremes one should begin to draw lines, where one should begin to question whether consent is truly informed, voluntary and therefore valid.[221]

This argument appears to support the claim that psychiatric assessments of decisional competence are more meticulous and nuanced than a presumption of FPCLC allows. However, in focusing on the risk of false positive determinations of competence, Thomas and Waluchow display scepticism of competence judgments, for which they do not provide supporting argument. That a legal standard of FPCLC fulfils a permissive function does not mean that it cannot be formulated on more restrictive grounds that recognises the high level at which the five general constitutive abilities of decisional competence must be pitched in order to offer a valid consent to biomedical research on one's own behalf. Second, Thomas and Waluchow, like Robertson earlier, appear to be using competence ambiguously. "Degrees... of competency" and "outright" incompetence are undefined, and could be taken to mean either decisional competence *or* FPCLC. Similarly, we do not know whether the reference to full competence and full incompetence refer to agency competence, task or decisional competence or a measure of the aggregate

---

[218]Council of Europe (2005, para. 7).

[219]Reich (1978: 333; 331) reproduced in Thomas and Waluchow (1998: 130; 132).

[220]Thomas and Waluchow (1998).

[221]Thomas and Waluchow (1998: 133).

task and decisional competences required for societal competence in North American society.

Suppose we set this aside and concede their argument relating to the problems of the legal standard. The logical outcome of their argument is the removal of competence from the realm of law to become the preserve of medicine or psychiatry and to be determined purely by the judgment of health care professionals rather than in accordance with an enforceable legal provision. Such thinking offers false promise. It will most likely result in elderly persons who have cognitive vulnerability receiving fewer enforceable protections when acting as research participants under codes of research ethics than they could under the law. An avoidance of or a dislike towards legal definitions of decisional competence does not imply that law cannot be used effectively to rectify the failings in its own previous attempts to define decisional competence.

Why Conduct Biomedical Research with Cognitively Vulnerable Groups?

Recent developments in neuroscience, such as neuroimaging, especially functional magnetic resonance imaging (fMRI), transcranial magnetic stimulation (TMS) and deep brain stimulation (many of which have been reported in the mainstream press in recent years) have reinvigorated widespread interest in biomedical research with cognitively vulnerable human participants, particularly research into the origins and treatment of mental disorder.[222] New drugs intended to manipulate neurochemical states for both treatment and 'enhancement' purposes are emerging at a rate that has led Farah et al. to term this trend "the psychopharmacopia of the early twenty-first century".[223] During the same period, there has also been a fuller appreciation of the prevalence of mental disorder within the population – a study published in 2005 reported that 27% of adults across the EU are experiencing or have experienced at least one form of mental health problem over the last twelve months.[224] This is shifting attention back towards the enhanced scientific knowledge obtained through research and the putative benefits for cognitively vulnerable groups that flow from this, rather than concentrating on its risks and the safeguards to which it must adhere.

---

[222] See Dana Alliance (2007), Glannon (2007), Malhi and Sachdev (2002), Farah (2002) and Farah et al. (2004 and 2005), in addition to the works cited at n. 18 above, for a discussion of recent developments. More specifically, see Faro and Mohamed (2006) on fMRI, George and Belmaker (2000) on TMS and DeLong and Wichmann (2001) on deep brain stimulation.

[223] Farah et al. (2004: 421).

[224] Wittchen and Jacobi (2005).

Degenerative mental disorders such as Alzheimer's disease,[225] which afflicts between five and six million citizens across the European Union[226] (and almost 700,000 in the UK alone[227]), variant Creutzfeldt-Jakob disease (vCJD, the human form of BSE),[228] which has been found to develop in adolescence, along with depressive and schizophrenic disorders[229], are conditions into which neuroscience has the potential to make significant contributions for understanding and treatment. Numerous research-council funded programmes rely upon the participation of individuals who themselves have been diagnosed as suffering from the conditions mentioned. This research, by its very nature, cannot be conducted with healthy volunteers.

Researchers are also collaborating between different areas of neurological study on an international basis. Trials conducted by the University of Pennsylvania with input from neurologists worldwide, published in 2003, claimed to establish a link between Alzheimer's disease and Parkinson's disease.[230] More generally, there has been progress made in understanding risk factors such as homocysteine, genetics, such as the suggested propensity gene apoE4, disease processes, pathology and causality, such as the 'cell cycle hypothesis'.[231] Such studies are set to multiply as the pace of scientific understanding quickens.

Patient support groups have also given their support to specific research programmes. For example, a clinical trial to investigate whether vitamin supplements can prevent or ameliorate Alzheimer's disease was conducted in 2002 by The University of Oxford in conjunction and with the imprimatur of the Alzheimer's Research Trust. According to the Trust, the proposed trial involved 3,600 elderly people with "mild cognitive impairment".[232] This degree of cognitive impairment inevitably bears upon the decisional competence that members of this group possess.

Many of these reasons for research share a common origin, namely the facilitation of medical understanding and a concern to develop treatments to offset the most destructive effects of such conditions. The scientific reasons for the participation of mentally disordered cognitively vulnerable individuals in biomedical

---

[225] Kim et al. (2001) mention in particular gene-transfer based therapy and cerebral call transplants. Farah et al. (2004 and 2005) and Simard and van Reekum (1999) discuss drugs for improving memory in dementia. Issa and Keyserlingk (2000) offer a thorough survey of research activity in the area and consider its ethical implications.

[226] Alzheimer Europe (2006).

[227] Alzheimer's Society (2007: ix).

[228] See Mead et al. (2007), Zerr and Poser (2002), Meikle (24/12/02) and BBC News Online (14/11/01).

[229] See, for example, Holtzheimer and Nemeroff (2006), Andreasen (2001) and Hoffman et al. (2000).

[230] Giasson et al. (2003).

[231] See, for example, Artiga et al. (1998) and Neugroschl and Davis (2002).

[232] Lavery (03/05/03).

research are set out in the Nuffield Council on Bioethics report, *Mental Disorders and Genetics, the Ethical Context*. These are:

1. Improved classification and diagnosis
2. Genetic counselling
3. Development of new and improved drug treatments
4. Improved preventive measures
5. Gene therapy[233]

The absence of any admission of therapeutic benefit to the research participant within the Nuffield Council's definition is indicative of the frequent 'non-therapeutic' value of such research, which is primarily anticipated to have long-term benefits to scientific understanding that will not directly benefit the research participant concerned, or at least certainly not in the way so-called 'therapeutic research' conventionally would. It is therefore important that the potential research participant is capable of understanding this and that she does not misperceive therapeutic aims.

The value of including children in medical research is evident from studies which focus on conditions unique to children as a developmental age group, or in the case of adolescents with mental health problems, to ascertain their responsiveness to for example, psychotropic medication which may need to administered in different quantities than with adult patients. This a view endorsed by the UK Medical Research Council in their 2004 ethics guidance Medical Research Involving Children.[234] More recently, this position has been enshrined across Europe within Directive 2001/20/EC on the implementation of good clinical practice in the conduct of clinical trials on medicinal products for human use.[235] Paragraph 3 states:

> … there is a need for clinical trials involving children to improve the treatment available to them. Children represent a vulnerable population with developmental, physiological and psychological differences from adults, which make age and development related research important for their benefit.[236]

The significance of the potential benefits medical research with children yields for their collective welfare creates a tension with strict consent-based requirements which inevitably lengthen the process of recruiting participants and conducting the research, thereby reducing the occurrence of such research projects. How one chooses to resolve this tension is connected with the value one attaches to consent relative to the other values at stake. The involvement of parents or carers in decision-making is one strategy designed to minimize the risk of medical research exploiting

---

[233]Nuffield Council on Bioethics (1998: 24). Gene therapy is a procedure where a gene is introduced into a cell so that there is a beneficial effect to the patient as a result of the action of the gene within the cell.

[234]Medical Research Council (2004: 7–8).

[235]Directive 2001/20/EC. OJ L 121, 1.5.2001, p. 35. All EU member states were obliged to implement this directive into their domestic legal systems by 2004.

[236]*Ibid.*

young people whilst ensuring that the child still can have some role in decisions about her participation.[237] Such practices of surrogate or shared decision-making raise ethical questions of their own.[238] However, in this book, my concern is with the grounds on which we should make a judgment of decisional competence to consent on an older child's own behalf where there is clear evidence to suggest decisional competence may be present. I do not intend to discuss how to proceed where it is reasonable to judge that a child does not have decisional competence to consent, as in the case with neonates, infants and young children, or not enough decisional ability to assume sole decision-making responsibilities.

Recognition that the neuroscientific community needs the participation of cognitively vulnerable individuals in biomedical research to facilitate understanding and innovation, particularly in respect of the aetiology of the conditions they experience, is a background upon which to appraise measures to regulate their involvement. Clearly, standards of decisional competence do not change with increasing scientific need and thus one should be wary about any attempt to relax standards of decisional competence for ulterior motives. Theorizing vulnerability allows us to recognise where cognitive vulnerabilities are more real than apparent, and how to go about dealing with them.

The discussions of vulnerability contained within the revised CIOMS Guidelines and the Explanatory Report are both encouraging, but they need to address definitions and standards of decisional competence to consent as part of a meaningful strategy for selecting research participants from cognitively vulnerable groups. To avoid reaching an impasse or erring in such situations, we must have recourse to moral reasoning that coherently resolves ethical questions raised when one is uncertain whether to judge a cognitively vulnerable person competent or incompetent to consent on their own behalf. In short, we need a moral strategy for resolving instances of doubtful competence.

[237] See Ross (1998) for a discussion.

[238] See Buchanan and Brock (1990, Chapter 5).

Chapter 3
Gewirth's Theory of Agency Rights

Moral philosophy is essential to conceptualising decisional competence in two ways. First, the substance of the particular moral theory to which one is committed determines how one chooses to understand the value of a judgment of decisional competence or incompetence. Second, to echo Jeffrie Murphy's observation in the introduction to this book, how one chooses to err in cases of doubtful competence – in favour of preserving decisional competence for as long as possible, or making a judgment of decisional incompetence at the earliest opportunity – is guided by the importance one's chosen moral theory attaches to maximising or minimising prospects for self-determination.

This chapter explains the structure of the Alan Gewirth's argument to the Principle of Generic Consistency (PGC) and the foundation of his theory of agency rights. I also consider the significance of consent and assistance in decision-making under the PGC, in order to explore the implications endorsing the PGC has for our understanding of competence and consent. This will facilitate my objective to present an argument from the PGC in relation to judgments of decisional competence in Chapters 4 and 5.[239] I end this chapter by illustrating how the PGC provides a cogent account of morality in its own right, by briefly contrasting its epistemological foundations with John Rawls's theory of justice as fairness and David Gauthier's rational contractarianism. I have chosen Rawls and Gauthier as comparators with the PGC because, like Gewirth, they both detail the steps of their arguments carefully and seek to provide an explicit foundation for moral action. Curiously, neither Rawls nor Gauthier address the argument to the PGC in their work, whereas Gewirth discusses the work of both philosophers.[240] However, I do not propose to

[239] Beyleveld (1991) distinguishes the argument *to* the PGC and arguments *from* the PGC in the following terms. The argument to the PGC consists of the arguments propounded in *Reason and Morality (RM)* that seek to demonstrate the dialectical necessity of the PGC from the internal viewpoint of any agent. Arguments from the PGC, alternatively, "are arguments applying the PGC to the resolution of familiar moral dilemmas". Beyleveld (1991: 398, n. 10).

[240] See *RM* (19–20, 108–109 and 340–341) for discussion of Rawls and Gewirth (1996: 11–12, n. 9) for discussion of Gauthier.

P. Bielby, *Competence and Vulnerability in Biomedical Research,*

analyse or refute counter-arguments to the PGC here, as Deryck Beyleveld has already undertaken this comprehensively.[241] The primary purpose of the chapter is to serve as an introduction to Gewirth's theory and the scope of its application, with particular reference to consent and decision-making.

The Argument to the PGC

In *Reason and Morality (RM)* and subsequent works,[242] Gewirth argues that two fundamental qualities characterize the action of any agent.[243] The ability of the agent to control her behaviour through her own volition whilst in possession of circumstantially relevant knowledge is voluntariness or freedom.[244] The reason an agent has for acting *at all*, which manifests in the agent's objective to fulfil a particular end, is purposiveness or intentionality.[245] The end in question could be either the action itself (e.g. dancing) or its consequences (e.g. to become more fit).[246]

Gewirth then relates this to the experience of a hypothetical moral agent. For any entity to qualify as an agent, he/she/it must be able to act for freely chosen purposes,[247] either currently or prospectively. This meshes with the account of agency competence proposed in Chapter 1. Agents need not necessarily be *human* beings – they could just as well be androids – as human beings do not exhaust the possible class of beings who could display the generic features of action.[248] For the purposes of this discussion, however, assume that an agent who is also human being – let us call her Elsa – is reflecting on her own action. Elsa is an individual in her own right, but she could represent *any* agent – including you or me – at anytime, anywhere, in any place. The steps of her reflection are contained in speech marks. When she performs an action, Elsa intends:

(i) "I do X for end or purpose E."

---

[241] Beyleveld (1991). See also, n. 21, above.

[242] *RM*: 22–198. The argument to the PGC as presented in this chapter closely follows the more concise structure presented in Gewirth (1982a, 1984a).

[243] Gewirth (1984a: 14).

[244] Gewirth (1984a: 14).

[245] Gewirth (1984a: 14).

[246] Gewirth (1984a: 14).

[247] *RM*: 44, Beyleveld (1991: xxxvi). By 'entity', I denote any human, animal, android or object that may conceivably possess agency. I prefer 'entity' to 'being' in order to avoid the association with 'having a life' or 'being alive', which is a contingent rather than a necessary feature of an agent's existence.

[248] This explains why we understand Gewirth's theory more accurately as a theory of agency rights (hence the title of this chapter), rather than a theory of human rights (Beyleveld, 1991: 447, see also Gewirth's acceptance of this in 1982a: 77).

X constitutes the means required in order to attain end or purpose E. Examples of this statement could be the reading of a book (X) in order to gain knowledge (E) or diving into a lake (X) in order to rescue a drowning child (E). Given the pursuit of E is the result of a freely expressed choice, Elsa attaches an importance to E such that she is motivated to act to bring it about.[249] Therefore, from *her* standpoint, this entails:

(ii) "E is good"

Two things are important here. First, the value that Elsa attaches to E does not necessarily need to be a moral value and will depend upon her particular choice.[250] Elsa may equally well intend to act to pursue a morally relevant end as much as a morally irrelevant end. However, common to any end or purpose E is the idea that Elsa must value the means employed to achieve E. Therefore, for her to will the end of her action, she must also will the means.[251] Second, in order for Elsa to succeed in achieving E, she must have the "proximate necessary conditions of action".[252] These proximate necessary conditions include the generic features of action broadened to include well-being.[253] According to Gewirth, well-being is:

> having the various substantive conditions and abilities, ranging from life and physical integrity to self-esteem and education, that are required if a person is to act either at all or with general chance of success in achieving the purposes for which he acts.[254]

Voluntariness, freedom, purposiveness, intentionality and well-being are the necessary conditions of *any* successful action, which Gewirth summarises as freedom and well-being.[255] Therefore, Elsa must hold:

(iii) "My freedom and well-being are necessary goods."

This means Elsa values freedom and well-being in a categorically instrumental way. Elsa may also express this as:

(iv) "I must have freedom and well-being."

'Must' in this context is a "practical-prescriptive requirement".[256] This means that it concerns what the agent is logically required to accept in order to avoid self-contradiction.[257]

---

[249] Gewirth (1984a: 14).

[250] Gewirth (1984a: 15).

[251] This is similar to Kant, who formulates means-ends rationality in similar terms: "Whoever wills the end, so far as reason has decisive influence on his action, wills also the indispensable necessary means to it that lie in his power" (quoted in Wood, 1999: 62).

[252] Gewirth (1984a: 15).

[253] Gewirth (1984a: 15).

[254] Gewirth (1984a: 15).

[255] Gewirth (1984a: 15).

[256] Gewirth (1984a: 15).

[257] *RM*: 193–195.

Mindful of this, Elsa acknowledges that:

(v) "I have rights to freedom and well-being."

Gewirth calls these rights 'generic rights' (GR).[258] But this, understandably, is a controversial move. In order to prove that (v) is the logical progression from (iv), imagine that Elsa rejects (v). Given the correlative duties of other individuals to act or refrain from acting in order to protect the object of the right to which Elsa makes a claim,[259] it follows that Elsa would also have to reject:

(vi) "All other agents ought at least to refrain from violating or eliminating my freedom and well-being."

In rejecting (vi), Elsa must hold:

(vii) "It is *not* the case that all other agents ought at least to refrain from violating or eliminating my freedom and well-being."

Elsa's commitment to (vii) entails:

(viii) "It is permissible for other agents to violate or eliminate my freedom and well-being."

So, by virtue of Elsa's commitment to (viii), Elsa must then hold:

(ix) "It is permissible that I may not have freedom and well-being."

However, it is clear that (ix) contradicts (iv), in which Elsa recognises the necessity of freedom and well-being for action.[260] In other words, Elsa (and all other agents) must reject (ix) because she must acknowledge (iv) as a condition of her acting for any freely chosen purpose *at all*. Therefore, acceptance of (v) is a matter of logical consistency for any agent.[261]

The rights claimed at this stage in the argument are prudential and not moral.[262] In order to elicit the transition from a prudential to a moral right claim, Gewirth appends several additional steps to the argument.

That Elsa is an agent, capable of both current and prospective purposivity, is the sufficient and necessary ground upon which she (and every other agent) affirms her rights to freedom and well-being.[263] Gewirth calls this "prospective purposive agency", abbreviated by Beyleveld to 'PPA'.[264] Elsa must therefore acknowledge:

---

[258] Gewirth (1984a: 16).

[259] Gewirth (1984a: 15). According to analytical jurist Hohfeld, human rights are primarily claim-rights. This idea is accepted by Gewirth in the context of the rights bestowed by the *PGC*. See Hohfeld (2001) and Gewirth (1978).

[260] Gewirth (1984a: 16).

[261] Gewirth (1984a: 16).

[262] Gewirth (1984a: 16).

[263] Gewirth (1984a: 16).

[264] Gewirth (1984a: 16).

(x) "I have rights to freedom and well-being simply because I am an agent (PPA)"[265]

We can demonstrate the validity of (x) by the Argument for the Sufficiency of Agency (ASA).[266] Imagine Elsa rejects (x) and instead insists that the only reason she has the generic rights of agency was due to the fact that she has blue eyes.[267] Consequently, Elsa would have to acknowledge:

(xi) "I have rights to freedom and well-being only because I have blue eyes"

In making this claim, Elsa is compelled to accept that, were it not for her having blue eyes, she would not have the generic rights. As such, Elsa contradicts herself.[268] This is because, if her eyes were a different colour, she would have to acknowledge:

(xii) "Because I do not have blue eyes, I do not have rights to freedom and well-being."

The acceptance of (xii) would conflict with the need for Elsa to hold necessarily that she has rights to freedom and well-being for no other reason than because she is an agent. Accordingly, Elsa must reject the notion that her having blue eyes is the sufficient and necessary justifying criterion of her having the generic rights and commit herself to (x).[269]

At this point, we move from the realm of prudential right claims to moral right claims. Now that Elsa has accepted (x), Elsa must also accept:

(xiii) "All agents have rights to freedom and well-being."

Gewirth derives (xiii) from (x) by a logical principle of universalization (LPU).[270] This requires, as a matter of logical consistency, the application of the PGC to all those other beings and entities who are agents. Gewirth explains this principle in the following terms:

> If some predicate *P* belongs to some subject *S* because *S* has a certain quality *Q* (where the 'because' is that of sufficient condition) then *P* must logically belong to all other subjects S_1 to S_n that also have *Q*.[271]

Elsa must therefore logically concede that all other agents have the generic rights for the same reason she has – simply because they, like her, are agents.[272] On that

---

[265] I use this abbreviation interchangeably with 'agent', although agent will be used predominately in this book to minimize the use of abbreviations.

[266] *RM*: 109–119, Beyleveld (1991: 43–45).

[267] This could be any contingent factor.

[268] Gewirth (1984a: 16).

[269] Gewirth (1984a: 17).

[270] Gewirth (1984a: 17), *RM*: 104–107. See also Beyleveld (1991: 44–45).

[271] Gewirth (1996: 18).

[272] Gewirth (1984a: 17).

basis, Elsa must accept that she should recognise and respect the generic rights of all other agents simply because, they, like her, are agents.[273]

Elsa now has a moral context for whatever action she envisaged in (i). This context is the agency rights claims of all other agents which her action may affect.[274] In order to avoid self-contradiction, Elsa along with all other agents are compelled to accept:

(xiv) "I ought to act in accord with the generic rights of my recipients as well as of myself."

We can restate this as the moral maxim of the PGC, which is to act in accord with the generic rights of one's recipients as well as of oneself.

It helps to clarify the structure of the argument to the PGC if we separate it into three stages.[275] The first consists of steps (i) to (iv). Here the agent must accept that she needs freedom and well-being in order to act for any freely chosen purpose. The second stage consists of steps (v) to (x). Here the agent is required to accept that she has a claim right to freedom and well-being (the generic rights), simply by virtue of being an agent. The third stage consists of steps (x) to (xiv). Here the agent must recognise out of logical consistency that every other agent, like herself, also has a claim right to freedom and well-being.

Gewirth calls this the 'Principle of Generic Consistency' because the argument unites logical consistency with the necessary conditions of any freely chosen action.[276]

Since Gewirth locates the sufficient and necessary justification of the existence the generic rights in the nature of free and purposive action, we may ascribe them to all beings or entities apparently capable of such action.[277] Although he conceives of the PGC as the justifying ground of *human* rights,[278] its responsibilities and protections clearly extend beyond human agents. As such, it is preferable to think of the generic rights as agency rights.[279] Insofar as the PGC obviates 'speciesist' objections levelled at conventional attempts to ground human rights as distinct from animal rights,[280] it is an egalitarian moral principle in the strongest sense.

The Methodology of the PGC

Gewirth refers to the philosophical method which the PGC uses as a *dialectically necessary* method, which is important to our correct understanding of it. The dialectical dimension of the method reveals itself in the "first person conative

[273] Gewirth (1984a: 17).

[274] Gewirth (1984a: 17).

[275] Beyleveld (1991: 13–14, 21–46).

[276] Gewirth (1984a: 17).

[277] The methodology I will use to identify of the moral community under the PGC is set out in Chapter 4.

[278] Most notably, in *RM* and in Gewirth (1982a), (1984a) and (1996).

[279] Beyleveld (1991: 447).

[280] See Singer (1995) for a discussion of speciesism.

perspective" of the agent, which is instantiated by Elsa's assertions (i)–(xiv) above, framed ("I do/I must/I need ..." etc.).[281] The dialectic nature of the method also allows for reflection upon the logical implications of the agent's statements.[282]

The dialectical *necessity* of the method derives from the fact that the PGC prescribes what all agents must logically claim and accept.[283] In other words, Gewirth draws the provisions of the PGC from the necessary claims of an agent. The opposite of this is an *assertoric* method, where the agent considers claims that she does not make from her own agential perspective to have the status of an objective truth.[284] Although the argument proceeds from the prudential rights-claims any agent must make, Gewirth argues that this does not compromise the force of the rights-claims or the categorical nature of the PGC.[285] According to Gewirth, "whatever is necessarily justified within the context of agency is also necessary for morality", which follows from the premise that agency constitutes the context of all moral action.[286] Correlatively, that which "logically must be accepted by every agent is necessarily justified within the context of agency".[287] This means that, notwithstanding the dialectically necessary method, we can describe the PGC assertorically.[288]

Gewirth offers two justifications for using the dialectical method.[289] First, "certain inferences that would not be valid apart from the conative first-person perspective of the agent *are* valid *within* that perspective".[290] To illustrate this, Gewirth analyses the terms in which he frames the argument to the PGC, contrasting the move from:

(i) "I do X for end or purpose E"

to

(ii) "E is good".

with:

(ia) "Some agent A does X for end or purpose E"

to

(ii) "E is good"[291]

Gewirth claims that (ii) does not follow from (ia) but (ii) does follow from (i).[292] We understand the difference between these two inferences that gives the former its

---

[281] Gewirth (1984a: 20).
[282] Gewirth (1984a: 20).
[283] Gewirth (1984a: 20–21), *RM*: 44.
[284] Gewirth (1984a: 21).
[285] Gewirth (1984a: 21).
[286] Gewirth (1984a: 21).
[287] Gewirth (1984a: 21).
[288] Gewirth (1984a: 21).
[289] *RM*: 42–47 and Gewirth (1984a: 21).
[290] Gewirth (1984a: 21).
[291] Gewirth (1984a: 21).
[292] Gewirth (1984a: 21).

validity if we recall the example of Elsa. In the inference from (i) to (ii), "E is good" is a claim made by Elsa (the agent herself) in the context her own purposive action.[293] Conversely, in the inference from (ia) to (ii), the claim "E is good" is made assertorically, as if it were being stated about an agent and her action by another person.[294] However, Gewirth points out that an agent's purpose is not good just because it is her chosen purpose – it could of course be very bad.[295] The important recognition captured by the use of the dialectically necessary method is that the agent considers her purpose to be good.[296] According to Gewirth, the proper meaning of (ii) is only conveyed if the agent herself makes the claim.[297] The "evaluative endorsement" of claim (ii) follows from the statement of action contained in (i) because the purpose in question is that of the individual agent which she accepts.[298]

Second, the dialectically necessary method "restricts the argument to what every agent is logically or rationally justified in claiming from within his conative relative standpoint in purposive action".[299] In taking a non-arbitrary starting point, Gewirth obviates accusations of speciousness or question-begging premises. A non-arbitrary starting point is one which logically binds all agents in action of some kind[300] – even that which seeks to bring about an end to one's agency, such as suicide. That there are logically irrefutable claims all agents must make about their action (irrespective of the *content* of that action) serves to ground the nature of action *per se*.[301] When the argument is followed through to its third stage (the move from self-interested to moral judgments), we are presented with a logically irrefutable benchmark against which to assess the moral acceptability of action.[302]

What dialectical necessity amounts to in practice is a heuristic for critical self-reflection whereby an agent is able to scrutinize the array of practical and moral judgments that she endorses, affirming those which are consistent with the PGC and rejecting those which are not.[303] This illustrates that all morally impermissible action is by definition rationally unjustifiable; it does not, as Gewirth observes, reduce ethics to logic.[304] From the claim that an agent requires the generic rights in order to engage in action of any kind, then, we are logically committed to accepting these tenets of the dialectically necessary method.

---

[293] Gewirth (1984a: 21).

[294] Gewirth (1984a: 21).

[295] Gewirth (1984a: 21).

[296] Gewirth (1984a: 21).

[297] Gewirth (1984a: 21).

[298] Gewirth (1984a: 22).

[299] Gewirth (1984a: 22).

[300] Gewirth (1984a: 22).

[301] Gewirth (1984a: 22).

[302] Gewirth (1984a: 22).

[303] Gewirth (1984a: 22), *RM*: 45–47.

[304] Gewirth (1984a: 22).

Defining the Content of Agency Rights

We saw earlier that in order for an agent to have the necessary means to pursue her freely chosen purposes, she must make a rights claim over the generic features of action, namely freedom and well-being. Although we can define freedom and well-being at a general level, they also have a specific content.[305] It is important to clarify them to understand the practical ramifications of Gewirth's argument.

For Gewirth, well-being is constitutive of three different elements. These are basic, non-subtractive and additive goods.[306] Interferences with these three types of good correspond respectively to three types of harm. Basic harm occurs when the right of an agent to basic well-being is infringed. Specific harm occurs when the right of an agent to non-subtractive or additive well-being is infringed.[307]

Basic goods represent the *sine qua non* of agency, and include "life, physical integrity and mental equilibrium".[308] An agent's right to basic goods is infringed when, amongst other things, she is killed, starved, tortured or unwillingly intoxicated.[309] These rights are further undermined where an agent is experiencing an infringement of their right to basic goods and another agent who could provide assistance without incurring similar hardships herself chooses not to do so.[310]

The second element of well-being, non-subtractive goods, "are the abilities and conditions required for maintaining undiminished one's level of purpose fulfilment and one's capabilities for particular actions".[311] Rights to these goods are infringed when the agent has her scope restricted for making plans and projects, when important information relevant to the context of her intended action is withheld or where some factor operates to frustrate the use of her own resources for the achievement of an end consistent with the PGC.[312] Lying, cheating, theft (assuming property arrangements that are themselves consistent with the PGC), deception and exploitation undermine rights to non-subtractive goods.[313]

The third element, additive goods, "are the abilities and conditions required for increasing one's level of purpose fulfilment and one's capabilities for particular actions".[314] Humiliation, denial of access to beneficial educational opportunities or discrimination on morally irrelevant grounds such as gender, sexuality, ethnicity, conscience, nationality or socio-economic background infringes rights to these

---

[305] Gewirth (1982a: 55).
[306] Gewirth (1982a: 55).
[307] Gewirth (1982a: 56).
[308] Gewirth (1982a: 55–56).
[309] Gewirth (1982a: 56).
[310] Gewirth (1982a: 56).
[311] Gewirth (1982a: 56).
[312] Gewirth (1982a: 56).
[313] Gewirth (1982a: 56).
[314] Gewirth (1982a: 56).

goods.[315] Actions that foster a climate of anxiety, mistrust or resentment (such as the curtailment of civil liberties) or activities that would ill-dispose agents to acting in accordance with the PGC, interfering with the development of wisdom, self-respect and other "self-regarding virtues", further infringe this right.[316] Gewirth identifies circumstances of ignorance, misunderstanding or superstition as liable to exacerbate this.[317]

Gewirth defines freedom as:

> a person's controlling his actions and his participation in transactions by his own unforced choice or consent and with the knowledge of relevant circumstances, so that his behaviour is neither compelled nor prevented by the actions of other persons.[318]

The converse of this – Gewirthian unfreedom – arises when an agent is exposed to violence of any kind, psychological or emotional abuse, coercion (of a type that is not permissible under the PGC), manipulation and deception "or any other procedures that restrict or remove his informed control of his behaviour by his own unforced choice".[319] The right to freedom under the PGC entails a right to autonomy and privacy which ensures that the agent may never have the will of others imposed on her unless it is with her fully informed consent.[320]

A Gewirthian moral wrong occurs when any action contravenes at least one of these dimensions of freedom and well-being.[321] The extent to which an action is morally wrong is dependent upon the degree of harm and the extent of encroachment on freedom that an agent would suffer were the action to occur. The infringement of the right of one agent by another agent is tantamount to accepting that a right claimed by the individual for herself insofar as she is an agent may be denied to another agent, despite the fact that she also has a claim to this right because she is an agent.[322] Immoral action is therefore an expression of irrationality. Whilst this does not necessarily foreclose ascriptions of responsibility for immoral action under the PGC, there are grounds for suggesting that if an agent consistently acts contrary to the PGC, or is incapable of guiding her action so that it conforms with the PGC, or infringes the PGC particularly gravely, then her societal competence diminishes and so too her level of responsibility.

The clearest absolute right, according to Gewirth, is the right of an innocent agent to life.[323] As such, it can be easily envisaged that there will be occasions on which the rights to freedom and well being will conflict. This could occur in three ways[324]:

---

[315] Gewirth only mentions three types of discrimination; however, it is clear that the PGC would accommodate a prohibition on all the additional forms of discrimination mentioned above.

[316] Gewirth (1982a: 56).

[317] Gewirth (1982a: 56).

[318] Gewirth (1982a: 56).

[319] Gewirth (1982a: 56–57).

[320] Gewirth (1982a: 57).

[321] Gewirth (1982a: 57).

[322] Gewirth (1982a: 57).

[323] Gewirth (1981).

[324] Gewirth (1982a: 57).

1. The freedom of agent A may conflict with the well-being of agent B when A uses his freedom to inflict some basic or specific harm against B.
2. The rights of different agents to well-being may conflict, as in the example where L must deceive M to prevent the torture of N.
3. The right of an agent to freedom and her right to well-being may conflict. Instances of this include suicide, sado-masochistic sexual practices or even excessive work when it interferes with one's health.

That we can envisage conflicts between competing agency rights suggests alternative approaches to their resolution. One way of doing this is to claim that the realisation of agency rights is conditional upon circumstance. This seems to be an unpromising response. It does not give us grounds for identifying which agency rights are engaged in which circumstances and what authority the circumstances have to attenuate their force. Another way that is more promising is to view rights as occupying levels of a hierarchy, similar to Ronald Dworkin's idea of 'rights as trumps'.[325] This admission, however, does not detract from the necessity of the agency rights themselves as guarantors of the generic needs of agency. It simply means that agency rights, framed as mere descriptions of what they are rather than accounting for the circumstances in which they are applied, do not generate their own criteria for resolution when two or more may be in conflict. Gewirth appends three grounds of resolution to his theory of agency rights, which he orders in terms of their importance for preserving agency.[326]

Resolving Conflicts of Rights Under the PGC

The first and most important strategy seeks to prevent or remove "transactional inconsistency", which circumscribes the right to freedom.[327] Where an agent intends to use her freedom in order to violate the freedom or well-being of other agents then the freedom of the violating agent can be curtailed in the relevant respect.[328] This curtailment follows from the maxim that the argument to the PGC yields – that each agent must act in accordance with the generic rights of all other agents.[329]

There are two main approaches to resolving transactional inconsistency, the applicability of which depends upon the nature of the inconsistency and the appropriateness of the responsive measure. The first is where agent A may coerce or harm another agent B in order to prevent B from coercing or harming either A or

[325] Dworkin (1984).

[326] Gewirth (1982a: 58).

[327] Gewirth (1982a: 58).

[328] Gewirth (1982a: 58).

[329] Gewirth (1982a: 58).

some other agent C.[330] The second approach justifies the imposition of state coercion designed to remove, avert or abate the original harm consistent with PGC-compliant legal rules, institutions and procedures.[331] In each case, the PGC requires that the nature and extent of the coercion or harm should be no more than is necessary for the purposes of resistance, removal or prevention. This amounts to a principle which states that in cases where there is an actual or intended infringement of one agent's generic rights by another, the PGC provides that the beleaguered agent or the state can take action in order to prevent this violation, provided the responsive measure is itself justified by the PGC.[332]

The second strategy looks to the degree to which the objects of the conflicting rights are necessary for the possibility of acting at all.[333] A more important right in this regard must be upheld over a lesser right where it is not possible to uphold one without infringing the other.[334] The example Gewirth uses to illustrate this is where agent A's right not to be deceived is overridden by agent B's right not to be murdered.[335] The third strategy bears similarities to this: one's right to freedom *may* be similarly restricted when the agent intends to infringe her own basic well-being, but only where there is doubt over her ability to fulfil the "emotional and cognitive conditions of freedom or voluntariness".[336] In these cases, we may use appropriate interventions, including reasonable force, in order to prevent an agent from causing unintended basic harm to herself. However, this paternalistic interference with an agent's freedom can only be licensed to prevent unintended basic harm, since the correlative basic goods are fundamental preconditions of action required if she is to maintain her status as an agent at all.[337] If one were to interfere with her freedom in order to prevent her causing harm to her own non-subtractive or additive well-being, then this would be unjustifiable under the PGC, unless her freedom was apparently compromised in the relevant respect. This is because these other two elements of her well-being are less central to her action than is her freedom.[338]

We can derive two conclusions from this. First, an agent intent on self-harm must ultimately be allowed to pursue her purpose if the emotional and cognitive prerequisites of voluntary action are met. Second, no-one has a right to restrict other agents from making choices that some agents may see as being misguidedly

[330] Gewirth (1982a: 58). Gewirth gives the following example: "If B physically assaults A or C, A may physically assault [i.e. use physical force against] B in order to resist or prevent the assault" (*ibid.*).

[331] Gewirth (1982a: 59).

[332] Gewirth (1982a: 58–59).

[333] Gewirth (1982a: 59).

[334] Gewirth (1982a: 59).

[335] Gewirth (1982a: 59).

[336] Gewirth (1982a: 59).

[337] Gewirth (1982a: 59).

[338] Gewirth (1982a: 59).

self-destructive, such as indolence or gluttony, where those actions do not impinge upon the generic rights of those other agents themselves.[339] As Gewirth observes:

> [The PGC] does not justify wholesale interference in the lives of others; it does not apply to projects that may lead only to decreased physical or mental efficiency... Persons must be left free to live their lives as they please and to make and perhaps profit from their own mistakes[340]

This leads us to the following distinction. Conflicts of rights can be resolved under the PGC where they exist between agents or in relation to one agent whose reflexive capacities are more likely to be seriously impaired. The PGC gives us no authority, however, to impose a resolution upon an agent where the conflict exists between her own rights and she has the decisional competence to waive the benefit of one of the affected rights. If cases of the latter type trouble us, it is more likely to be because we are ourselves uncertain about the decisional competence of this agent, rather than by what her choices entail.

Direct and Indirect Applications of the PGC

Before we establish the viability of an argument from the PGC, we need to consider the modalities of its application. These are direct and indirect methods.[341] The direct application of the PGC places a requirement that the actions of each agent are in conformity with what is morally permissible under the PGC.[342] The indirect application of the PGC requires the creation of PGC-compliant social rules and institutions. In order for these rules and institutions to be valid, they must uphold the equality of freedom and well-being of all agents.[343] Any agent who acts in such a way that these rules and institutions permit will by definition discharge their moral duties towards other agents.[344] As we have seen, the indirect application may provide for agents to be coerced *without* infringing their rights to freedom and well-being, in such cases where the PGC justifies coercive rules and institutions in order to protect the generic rights of other agents, and in the case of apparently involuntary self-harm, the agent herself.[345]

---

[339] We should remember that whilst agents are free to choose purposes of this nature, the potential social costs incurred warrant extensive educational initiatives that would offer those individuals greater information upon which to decide whether to pursue more conscientiously their additive well-being. See *RM*: 240–248 and 265.

[340] *RM*: 265.

[341] Gewirth (1982a: 60).

[342] Gewirth (1982a: 60).

[343] Gewirth (1982a: 60).

[344] Gewirth (1982a: 60).

[345] Gewirth (1982a: 60). An example of PGC-compliant coercion is the indefinite detention of a psychopathic serial killer in a psychiatric hospital or other therapeutic environment.

Gewirth subdivides the indirect applications of the PGC into two distinct forms. Procedural applications ensure the moral justifiability of social rules and institutions which are the product of agents' consent.[346] These follow from the PGC's inherent concern with freedom and resemble arguments present in established social contractarian theories.[347] According to Gewirth, procedural applications have optional or necessary dimensions. Optional procedural applications pertain to the consent required for voluntary associations.[348] Necessary procedural applications pertain to the consent that is required as part of the decision-making that mandates democratic bodies and their legislative powers, such as elections for political representatives.[349]

Instrumental applications, on the other hand, ensure the moral justifiability of social rules and institutions which uphold the well-being of agents and follow from the PGC's inherent concern with well-being.[350] Instrumental applications of the PGC take either a static or a dynamic form.[351] Static instrumental applications protect agents from infractions of their generic rights and sanction those who violate these rights.[352] The closest approximation to this in existing institutional arrangements is the criminal law, although this does not mean that the PGC would endorse the array of punitive sanctions that such systems commonly employ.[353] Dynamic instrumental applications uphold access to and preservation of basic, non-subtractive and certain additive goods over time that the agent cannot achieve alone.[354] An expansive and properly resourced welfare state, which ensures at the very least equality of opportunity in the conditions necessary for successful action, embodies these protections.

Although the PGC does clearly prescribe what must be fulfilled both by the actions of individuals and by the role of the state, legal enforcement of all agency rights is not necessary.[355] However, all legally enforceable rights justified by the

---

[346] Gewirth (1982a: 61).

[347] Gewirth (1982a: 61). For a comprehensive overview of such theories, see Boucher and Kelly (1994).

[348] Gewirth (1982a: 61).

[349] Gewirth (1982a: 61).

[350] Gewirth (1982a: 61).

[351] Gewirth (1982a: 61).

[352] Gewirth (1982a: 61).

[353] The PGC does have an important compassionate dimension manifest in both its interpersonal and institutional application (Gewirth, 1996: xv, 21–22, 83, 1998: 71, 87). This extends to inform Gewirthian criminal justice ethics. Brown (1998) proposes a theory of punishment derived from the PGC, though does not intend for this to specify what the precise content of the sanction would be in certain types of cases. All things considered, it is highly likely that any criminal justice system founded on the PGC would favour restorative justice practices and rehabilitation rather than punitive incarceration, because of the primacy of well-being and the absence of any place in the PGC for vengefulness. Commentators tend to overlook this dimension of the PGC, given the fixation with the structure of the argument rather than its application.

[354] Gewirth (1982a: 61).

[355] Gewirth (1982a: 61).

PGC are species of agency rights. This amounts to a claim that, in a PGC-compliant polity, not all moral rights should become legal rights but that all legal rights are moral rights. Those rights that should receive legal protection in a PGC-compliant polity are those whose breach has a seriously impact upon the agent's generic goods (e.g. bodily and psychological integrity, education, privacy) but not those whose violation results in a minimal impact on the agent's interests (e.g. there can be no right against trivial promise breaking).[356]

Using the framework delineated above, Gewirth identifies three different methods to enshrine in law the applications of the PGC that warrant legal protection[357]:

1. The *static-instrumental* justification of legal protection.[358] This ensures fundamental agency rights are not violated by other agents, corporate entities and the state and is manifest in the coercive or prohibitive aspects of the law (such as the criminal law).[359]
2. The *dynamic-instrumental* justification of legal protection.[360] It is a fact of life that agents are, through no fault of their own, positioned unequally in terms of their ability to secure and maintain their generic rights. The rules that emanate from this justification seek to dissipate such inequalities, through the provision of healthcare, education, housing and monetary benefits on behalf of the state to all who need them.[361] The dynamic-instrumental protection also comprises the regulatory function of the state to ensure that the standards of services and utilities are maintained at a level commensurate with respect for the rights of all agents under the PGC.[362]
3. The *necessary-procedural* justification of legal protection.[363] This requires that all laws, officials and governmental action are informed by consensual procedures and comprises civil liberties enshrined through constitutional provisions and institutional design.[364] The significance of this is extensive. The protection provides for fundamental civil liberties such as freedom of association, movement, assembly, expression and political participation, and a form of social libertarianism whereby the state refrains from interfering in the freely chosen actions of any agent provided that they do not interfere with the generic rights of others.[365] A vast sphere of social activity should therefore be beyond the reach of official regulation and the right to engage in these activities should receive state protection.[366]

---

[356] Gewirth (1982a: 61).

[357] Gewirth (1982a: 61–62).

[358] Gewirth (1982a: 61–62).

[359] Gewirth (1982a: 61–62).

[360] Gewirth (1982a: 62).

[361] Gewirth (1982a: 62).

[362] Gewirth (1982a: 62).

[363] Gewirth (1982a: 62).

[364] Gewirth (1982a: 62). Gewirth refers to this as "the methods of consent".

[365] Gewirth (1982a: 63).

[366] Gewirth (1982a: 63).

When applying the PGC to issues of decisional competence, and the moral justification for ascribing FPCLC to consent or to refuse research participation, it is the dynamic-instrumental and necessary-procedural justifications of legal protection with which we are primarily concerned. The levels of education and information necessary to make such a choice can only be provided by state action that is designed to support individual autonomy, which goes to the first of these justifications. Whether the state has a moral justification to interfere with the content of an individual's choice is a matter of civil liberties and goes to the second of these justifications.

Consent and the PGC

Consent serves a twofold function under the PGC. First, it protects the individual from an unwilled interference in her life where there is no overriding human rights-based justification for doing so. Second, the right of choice that consent offers gives expression to human dignity,[367] as being a dignity-holder is derivative from being an agent.[368] As Brownsword acknowledges, consent is not itself a human right but instead "parasitic upon" a morally prior framework of rights and duties.[369] The role of consent is as a procedural justification to grant or withhold authority for interference with the object of the right (e.g. not to have one's bodily or psychological integrity interfered with) where no over-riding human rights-based justification is engaged.[370] Consent is therefore a process that legitimises the waiver of the benefits of the rights at stake on the sufficient and necessary condition that (a) the individual concerned can understand the full implications of waiver and (b) this does not jeopardize the rights of other agents. The placing of consent as a procedural value in Gewirthian theory allows us to avoid the pitfalls of viewing it as an end in itself, which as Brownsword argues, can lead to a problematic fixation with consent.[371]

In the context of biomedical research, consent is invoked at the level of defining biomedical research with human participants and gives it its ethical character.[372] Consent performs a similar function in defining biomedical research as it does in forming part of the definition of sexual intercourse, without which the definition of the act would become that of another, namely rape. The implicit presence of consent in the definition of biomedical research separates it from a notion of physical or

367 Beyleveld and Brownsword (2001: 242).

368 Gewirth (1998: 208).

369 Brownsword (2004: 229).

370 Brownsword (2004: 225, 228). Beyleveld and Brownsword (2007a: 335–336).

371 Brownsword (2004).

372 On the inseparability of moral judgments from concept formation in the social sciences more generally, see Toddington (1993).

psychological violation in the name of medical progress.[373] Unlike sexual intercourse, however, the scope of consent in biomedical research is not limited to the person who is to participate (i.e. first-person contemporaneous legal competence (FPCLC) consent), but extends to proxy consent if the potential participant is decisionally incompetent.

Nonetheless, the PGC goes much further than simply justifying consent as the conduit for the legitimate waiver of a negative right. The generation of a positive rights claim from the structure of the argument means that consent – as a procedural justification behind the modification or waiver of a substantive right – places a duty of assistance on the part of others to help the person understand the implications of waiving the benefit of that right. This is a corollary of the positive rights-claim to the object of the right itself. It follows from the PGC-protected right to have knowledge of circumstances relevant to the particular context of action.[374] To deny this amounts to a breach of the ethical duty that follows from the positive dimension to the right. This duty of assistance requires responsiveness to the needs of the agent in question if we are to make a sincere effort to assist her to understand, even if it transpires that she apparently cannot understand, due to decisional incompetence. Common forms of assistance include education and information provision that is appropriate to the developmental stage that the person has reached. This makes the processes leading up to offering or withholding consent sensitive to the psychological needs and dispositions that accompany the experience of cognitive vulnerability in particular and of baseline vulnerability in general.

The consequences of this for consent in biomedical research are wide-ranging. There is an ethical duty incumbent upon anyone undertaking research and those responsible for its oversight to be mindful of the cognitive and/or circumstantial vulnerabilities of the individuals approached to participate.[375] The ethos therefore shifts from obtaining consent to empowering the potential participant to decide.[376] On a practical level, this involves putting in place mechanisms to assist actively the potential participant's understanding and appreciation of the research process. As a minimum, this includes explanation of its objectives; the nature and consequences of participation and the rationale upon which she is being approached to have her consent sought in the first place. Where an individual has a questionable ability to make decisions about participation, further and more specialised assistance is warranted, ideally from someone who does not have a direct interest in the research going ahead.

---

[373] Bielby (2005a: 222). This ethos underpins consent provisions in the earliest codes of research ethics of the post-Second World War period, such as the Nuremberg Code and the World Medical Association Declaration of Helsinki.

[374] Gewirth (1978: 250–52, 258, 260).

[375] This should not be seen as mere compliance with 'best practice' in ethical review, but amounts to a duty to seek out new ways to be responsive to these vulnerabilities as far as possible.

[376] See McMillan and Gillett (2002: 225) for a discussion of the empowering potential of consent and Feenan (1997) for a discussion of empowerment in relation to assessment of competence to consent to treatment.

The case of individuals with cognitive vulnerability epitomises the importance of appropriate assistance. For them, the mere provision of simplified consent forms, greater explanation of research procedures, or provision of information in alternative formats alone, may not offer an improved decision-making situation relevant to their needs. If we accept the importance that Gewirthian theory attaches to the justification of consent and the way in which it should be sought, then we have a reason for preserving the decisional competence of potential research participants as far as possible.

On a more theoretical level, the Gewirthian view on consent moves the bioethical debate away from unhelpfully bifurcated thinking about autonomy and paternalism. Instead, it recognises that individuals should receive assistance in making decisions for themselves and that this is not something that is likely to happen without active interpersonal support. Such interventions are best articulated as duties which attach to particular roles (e.g. physician, researcher or counsellor), although we can also imagine them arising in everyday contexts that do not involve the seeking of consent, such as where a person reads out the bus times to a partially sighted person in order to help her plan her journey. These represent interventions that seek to promote the autonomy interests of the individual concerned, and elicit her decisional independence, motivated by a sincere concern for her dignity and capability as an agent. We will return to this in Chapters 5 and 6.

An Evaluation of the PGC Against Two Alternative Rationalist Ethical Theories

John Rawls

One of the most well-known and widely debated theories to emerge in moral and political philosophy during the last century is in the work of John Rawls.[377] In *A Theory of Justice*,[378] Rawls established what he considered the integral elements of a theory of justice as fairness. Justice in this sense is only possible if social institutions do not operate to allow individuals to benefit from talents and endowments arbitrarily bestowed at birth, unless in doing so, the exercise of these talents benefits the worst off.

[377] In this section, my concern is with the 'early' Rawls of *A Theory of Justice* (1972, rev. ed. 1999) rather than the 'later' Rawls of *Political Liberalism* (1993). In *A Theory of Justice,* Rawls undertakes an attempt to construct a moral theory of rights (his account of 'justice as fairness'), as contrasted with the concern of *Political Liberalism,* which is to theorize justice as fairness in a way that is (supposedly) acceptable to all citizens in a democracy (the idea of an 'overlapping consensus'). Insofar as Gewirth's aim in *Reason and Morality* is to argue for a rights-based moral epistemology, *A Theory of Justice* is therefore the better comparator.

[378] Rawls (1972, rev. ed. 1999).

The central stages of Rawls's account take the following form. In order to create the framework for a just society, we must first construct a thought experiment. We should imagine ourselves in the 'original position' – a position of strict equality in which nobody has antecedent knowledge of her eventual human traits (e.g. gender, ethnicity, sexuality, religion, intelligence (dis)abilities, etc.). This, Rawls claims, is known as the 'veil of ignorance',[379] which occludes all foresight of these socially significant facts. Behind this 'veil', we are to construct principles of justice that would benefit all to an equal extent, irrespective of whatever human form or situation we actually come to occupy. In this way, Rawls believes, we are motivated to adopt the same concern for the fate of everyone in society.

The principles we would devise under this fictitious ignorance amount to a hierarchy of importance or a "lexical order of priority",[380] with the first taking precedence over the second, the second over the third and so forth. Rawls states the content of these principles as follows:

(i) Each person is to have a right to the greatest equal share of liberties compatible with a similar right for all.
(ii) (a) Social and economic inequalities are to be attached to offices and positions open to all according to fair equality of opportunity.
   (b) Such inequalities are justified if and only if they benefit the least advantaged in society. Rawls refers to this as the "difference principle".[381]

The first principle is commensurate with the notion of equal liberty. This means that everyone would be free from prejudice, oppression or persecution of any kind. The second principle prescribes that equality of opportunity should allow anyone with sufficient skill and ability to flourish, irrespective of socio-economic background. The difference principle (iib) affects the principle of the distribution of inequalities (iia) as it provides for differential outcomes only insofar as this improves the position of the most deprived.

Rawls's theory of justice is essentially an attempt to use rational choice theory as the first principle of liberal-egalitarian thinking.[382] When placed in the 'original position' of equality in respect of the distribution of freedom, power and wealth, and unaware of one's human characteristics when concealed by 'veil of ignorance', Rawls believes that individuals will choose this egalitarian moral principle as a result of applying rational choice processes. However, as Gewirth identifies, there are two ways in which Rawls's attempt to justify this moral principle fails, both of which the argument PGC avoids.[383]

The first flaw arises in the inability to appeal to "independent rational justification" (using the canons of inductive and deductive logic) to endorse the use of the veil of

---

[379] Rawls (1999: 118–123).

[380] Rawls (1999: 37–38).

[381] Rawls (1999: 266–267).

[382] *RM*: 19.

[383] *RM*: 19–20.

ignorance or the original position.[384] In as much as Rawls is correct about the unequal distribution of abilities, talents and endowments, it follows that individuals are not similarly situated to undertake the thought experiment that the Rawlsian project requires of them. Even if we were to concede to Rawls that reasoning in the original position behind the veil of ignorance is a capacity exercisable by all, a problem remains. The demands that the processes of abstraction place on individuals, particularly the veil of ignorance, surpass that minimal level of ignorance that rational persons commonly accept when making choices under uncertain conditions.[385] It is also questionable whether individuals could sufficiently alienate themselves from knowledge of their own selfhood (even if only temporarily) in order to eschew fully *all* their contingent human qualities, particularly the most embedded ones such as character and temperament, that comprise their existential position.[386]

One might counter that Rawls's argument is persuasive independent of whether an individual chooses to understand it or has the capacity to follow it. However, the methodology Rawls uses generates its own limitations on how far this line of defence extends. The methodology involves a dialectically contingent process that requires evaluation on its own merits.[387] Unlike Gewirth's argument to the PGC, nothing about it is rationally ineluctable. To claim that the argument is forceful presupposes that the person making this claim considers it to represent a persuasive basis on which to ground moral principles of justice. This inevitably requires the philosophical skill to scrutinize the argument and the intellectual capacity to understand counterfactual reasoning. The methodology also rests upon an understanding that risk-averseness is the most appropriate basis on which to go about decision-making under conditions of uncertainty. This is a contingent rather than a necessary claim that invites controversy.

The second flaw emerges in the criterion for arranging the principles of justice, which Rawls terms the "lexical order of priority". Rawls defines this as:

> [A]n order which requires us to satisfy the first principle in the ordering before we can move on to the second, the second before we consider the third, and so on. .. [This] avoids … having to balance principles at all; those earlier in the ordering have an absolute weight, so to speak, with respect to later ones, and hold without exception.[388]

Whereas the criteria for resolving rights claims under the PGC is derived from a hierarchy of goods whose importance for action determines their weighting, Rawls's lexical order is not strictly determined by the principles of justice but instead by external factors that are separate from the principles.[389] This amounts to Rawls's own presumption about the motives that would influence people to choose particular principles of justice over others in the original position behind the veil of

---

[384] *RM*: 19.

[385] *RM*: 20.

[386] For a discussion of this point, see Sandel (1998).

[387] Gewirth (1996: 27).

[388] Rawls (1999: 38).

[389] *RM*: 340–341.

ignorance.[390] These motives themselves have no ineluctable rational foundation, unlike the motive of an agent to recognise her claim to the generic rights under the PGC. By comparison, Rawls's lexical ordering appears rather arbitrary, exhibiting, as Gewirth observes, the particular concerns of established liberal political philosophy without offering a rational justification for them.[391]

As such, whilst Rawls's argument may be valid within its own terms of reference, it ultimately represents an arbitrary basis upon which to justify moral principles. Gewirth's argument to the PGC, alternatively, has premises grounded in the basic features of action that are dialectically necessary and rationally ineluctable for any agent. Although the argument itself operates at a similarly high level of abstraction, and requires the capacity for sustained logical reasoning if one is to follow it successfully, the PGC does not require that the agent understand the argument – or even agree with it – in order for it to have force. In other words, acceptance of the PGC is logically necessary irrespective of my capacity to know that this is the case, or my keenness to dispute this.[392] This does not apply to Rawls's theory of justice.

Rawls does present a meticulous and attractive proposal about how individuals can devise social and political arrangements to discharge their moral obligations along rational lines. However, this only provides us with a rational way to identify the individuals who are the recipients of our moral obligations and the interests of which we must take favourable account when discharging our moral obligations towards those individuals. What Rawls does not do is to provide a rationally compelling answer as to why one should be moral at all.[393] In short, Rawls provides a rational way to be moral but does not explain why rationality *entails* morality.

David Gauthier

The moral contractarianism of David Gauthier is quite distinct from the liberal egalitarianism of either Rawls or Gewirth. In *Morals by Agreement*,[394] Gauthier elaborates a basis for morality founded upon principles of rational choice. For Gauthier, rational choice represents the constrained maximization of self-interest.[395] This notion of rational choice relies upon a distinction between two main types of choice situation. 'Parametric choice' situations are fixed choice scenarios where the rationality of the choice derives from the greatest anticipated utility.[396] 'Strategic

[390] *Ibid.*: 341.

[391] *Ibid.* Gewirth observes how this is particularly evident in the prioritising of individual liberty over economic security.

[392] See Beyleveld (1991: 149–150, 305–306, 477).

[393] Beyleveld (1991: 310–311).

[394] Gauthier (1986). The critique presented here draws upon Vallentyne (1991).

[395] Vallentyne (1991: 2, 5–6).

[396] Vallentyne (1991: 6), Gauthier (1986: 21, 85, 170–171, 183–184). By 'utility', Gauthier means the most effective manner through which the agent can achieve her goals.

choice' situations, alternatively, are those where a choice's rational value derives partly from what other individuals would rationally seek to choose.[397]

However, this is where any similarity with Rawls's theory ends. Gauthier defines a person as someone who selects what is likely to offer the greatest expectation of value or utility to herself.[398] In doing so, Gauthier relies upon a highly egoistic model of human personality and motivation. This is reflected in how he explains the move from human isolation to co-operation. In the absence of conditions of perfect competition, any agent who maximizes her interests without constraint will undoubtedly disadvantage themselves and others.[399] Morality functions so as to place rational limits upon self-interest to ensure that everyone can benefit equally through the actions of each other.[400] Nevertheless, Gauthier is not seeking to devise a moral theory as such, but a theory of rational action. In particular, his concern is to justify limitations on human behaviour that are both rational and impartial.[401] Indeed, Vallentyne remarks of Gauthier that, "[i]t is not merely that his theory might fail to capture some traditional moral concerns, but rather that its connection with these traditional moral concerns is purely contingent."[402]

Gauthier's theory begins with an 'initial bargaining position', a fiction in which individuals neither interact nor co-operate, subject to the Lockean Proviso.[403] In this position, no interaction of any kind takes place, a manifestly unsustainable state of affairs that benefits no-one.[404] This impasse can be overcome, Gauthier argues, through adopting a co-operative bargaining position of constrained maximization that is ultimately more desirable from the perspective of long-term self-interest. This involves employing a strategy that will minimize anyone's maximum relative concession to the interests of others.[405] Following Vallentyne, we can express an individual's relative concession for a particular option in ratio form:

(a) "the excess of (i) the utility for that person of his/her most favourable admissible option over (ii) the utility for that person of the given option

to:

(b) the excess of (i) the utility for that person of his/her most favourable admissible option over (ii) the utility for that person of the initial bargaining position option".[406]

---

[397] Vallentyne (1991: 6), Gauthier (1986: 21, 24, 61, 68, 76–78, 157–158).

[398] Gauthier (1986: 9).

[399] Vallentyne (1991: 2) and Gauthier (1986: 2). This is epitomized by the tale of the Prisoner's Dilemma.

[400] Vallentyne (1991: 2), Gauthier (1986: 9, 11).

[401] Vallentyne (1991: 2).

[402] Vallentyne (1991: 2).

[403] Vallentyne (1991: 7). This is an idea derived from the political writings of John Locke that no-one should profit when in doing so someone else would experience a loss. The term originates in Nozick (1974: 175–182).

[404] Vallentyne (1991: 7).

[405] Vallentyne (1991: 8).

[406] Vallentyne (1991: 8) and Gauthier (1986: 136–148).

Any possible option must offer every party to the co-operative bargaining position as a minimum the same utility as in the initial bargaining position, and must be realisable.[407] Since a choice's rational permissibility is governed by its conformity with a rationally justifiable policy,[408] Gauthier argues that constrained maximization is rational as it will ultimately maximize our self-interest.[409] However, this arrangement will only function, according to Gauthier, provided we honour the rational agreements that we have made with others who are themselves disposed to honouring their rational agreements. Exclusion from the benefits of bargaining arrangements is the almost certain response if we were not to comply, as other rational agents would have no grounds to depend upon our integrity.[410] It is therefore in our own interests to adopt a character that is trustworthy and reliable and a policy, known as 'choice disposition', of complying with the terms of rational agreements.[411] This contrasts with the ultimately disadvantageous position we would find ourselves in if we contrive a trustworthy and reliable character whilst in fact disregarding compliance with rational agreements.[412]

Gauthier seeks to proceed from a starting point of moral neutrality in applying rational choice theory.[413] Insofar as Gewirth also begins from the neutral starting point of fundamental requirements for action, there is an ostensible affinity between Gauthier's contractarianism and the PGC. Nevertheless, this is where such affinities begin and end. Because Gauthier seeks to ground morality in rational agreement, he maintains that such agreement requires mutual advantage.[414] Gauthier defines the participants in the agreement as those individuals who are living and potential contributors to a co-operative arrangement.[415] As such, only these individuals have intrinsic moral status. For Gauthier, children, adults with profound intellectual or physical disabilities, animals and future generations may not enter into the agreement because they fail to offer benefits to the participants that they cannot secure already themselves.[416] These groups only have derivative moral status in the sense that they are of value to participants in the agreement.[417] Applying this principle to health care, Gauthier claims:

> From a technology that made it possible for an ever-increasing proportion of persons to increase the average level of well-being, our society is passing to a technology, best exemplified by developments in medicine, that make possible an ever-increasing transfer of

[407] Vallentyne (1991: 8).

[408] Vallentyne (1991: 10).

[409] Gauthier (1986: 167–170).

[410] Vallentyne (1991: 10).

[411] Vallentyne (1991: 10), Gauthier (1986: 182–184, 186–187).

[412] Gauthier (1986: 173).

[413] Vallentyne (1991: 2–3).

[414] Vallentyne (1991: 4).

[415] Vallentyne (1991: 4).

[416] Vallentyne (1991: 4), Gauthier (1986: 268).

[417] Vallentyne (1991: 4).

> benefits to persons who decrease that average. Such persons are not party to the moral relationships grounded by a contractarian theory.[418]

Gauthier's adoption of a social Darwinist position towards those who impede the initial bargaining position rather than one which would help disadvantaged groups pursue their interests and develop their autonomy, demonstrates that his theory is founded upon mutual unconcern, to an extent that many would find intuitively disturbing.[419]

Even if Gauthier's uncompromising position on contractarian moral standing is set aside, the question still remains why constrained self-maximisation should ground moral principles. The identification of natural inequality amongst persons in Gauthier's moral contractarianism, as in Rawls's justice as fairness, represents a sensible starting point.[420] However, following Gewirth, I believe that Gauthier fails to achieve the impartiality that he claims for his theory,[421] even in spite of his attempts to defend it.[422]

The problem arises when we begin to analyse why those who benefit from the distribution of natural inequalities should choose to enter into an impartial bargain with those who are disadvantaged by the distribution at all.[423] Although disadvantaged individuals may have something to offer those who are advantaged, at a cost the naturally advantaged can afford, this itself does not constitute a necessary let alone a sufficient reason why individuals enter into bargaining arrangements.[424] To accept this would commit one to a contingent and narrow view of human motivation, as opposed to the universal premise from which Gewirth starts, the need for freedom and well-being if one is to act with any chance of success at all. Gauthier appears to make the mistake of much classical liberalism insofar as he appears to conflate rationality with idealised decision-making in market economics and the model of the self it presupposes. Gauthier may purport to offer a rational explanation for morality (albeit a narrow one), but like Rawls, does not answer our question of *why* it is rational to be moral in the first place.

Summary

Gewirth's theory of agency rights provides a rationally ineluctable theory of morality which is compelling both as a purely abstract ethical principle and as the foundations of a political and bioethical theory which stresses universality, egalitarianism and

---

[418] Gauthier (1986: 18).

[419] Even though, Gauthier would retort, there is no place in this theory for moral intuitionism (1986: 269; Vallentyne, 1991: 2). Of course, the same applies to the PGC, although it gives rise to far fewer counter-intuitive moral conclusions.

[420] Gewirth (1996: 11–12, n. 9).

[421] Gewirth (1996: 11–12, n. 9).

[422] Gauthier (1988, 1991).

[423] Gewirth (1996: 11–12, n. 9).

[424] Gewirth (1996: 11–12, n. 9).

care. In addition to providing an argument for agency rights, it also recognises the inherent dependence and vulnerability of all agents and suggests lines along which institutions and social practices could be designed in order to offer support for all agents in need of it. This may be countered as another visionary 'grand narrative', but if the argument to the PGC is accepted, then it is not possible to deny the existence of the generic rights without contradicting what I am implicitly committing myself to in the act of that denial.

Following Beyleveld and Brownsword, the proper place of consent is Gewirthian theory is as an important procedural safeguard of rights to bodily and psychological integrity.[425] Consent may not constitute a substantive right in itself, but its ethical and legal value derives from its ability to create and modify relationships which the generic rights circumscribe. It follows that, in a PGC-compliant polity at least, all individuals require appropriate forms of education and empowerment to be able to engage in decision-making to the extent of their abilities. Difficulties arise when decision-making abilities are merely assumed to exist in an individual without an attempt to consider whether some decision-makers will require support to elicit their decision-making abilities. That some agents experience cognitive vulnerability in ways that could undermine their decisional competences illustrates these difficulties. Practices that support individual decision-making epitomize the positive rights to assistance under the PGC.

There is nothing in either Rawls or Gauthier's theory which serves to undermine the argument to the PGC nor to offer a more compelling approach to the justification of morality. The comparison with Rawls and Gauthier has illustrated the superiority of grounding morality in the necessary features of action, rather than through screening out morally irrelevant characteristics of human existence as in Rawls, or presupposing that all human interaction is conducted from the perspective of self-interest, as in Gauthier.

One might object that to choose two theories of a similar kind with which to compare the PGC is selective and negates approaches generated from postmodernism, feminist theory and non-Western philosophy.[426] To analyse the PGC in terms of the vast scholarship in this area would require a book in itself and, for present purposes, would not necessarily make Gewirth's argument for agency rights any clearer than it has been presented here. As subsequent chapters will show, I do not wish to dismiss alternative insights but instead to draw upon them where appropriate in developing a theory of competence judgment that is consistent with the tenets of the PGC. Mindful of the ways in which it operates, the next chapter sets out the precautionary basis on which we must apply the PGC and on which the moral defensibility of judgments about decisional competence depend.

---

[425] Beyleveld and Brownsword (2007a, Chapter 11).

[426] For a discussion of these approaches as applied to bioethical issues, see Shildrick (1997), Wolf (1996) and Alora and Lumitao (2001) respectively.

Chapter 4
Proportionality, Precaution and Judgments of Competence

It is one thing to endorse a theory of morality; it is quite another to be clear about how we come to identify members of the moral community, and the moral status of the entities within it.[427] By itself, a definition of agency cannot offer an argument for making ascriptions of agency competence, let alone to identify whether an agent has the ability or potential to develop task and decisional competences. Devising such an argument may seem like a straightforward application of the theory to the empirical world. However, on closer examination, matters are not so simple.

The very idea of a moral community raises a fundamental question of moral epistemology. All moral theories, including the PGC, need to answer the claims of the sceptic as to how we can know that entities worthy of moral consideration exist. That there are other such entities is a proposition many moral theorists are frequently too hasty to accept and too reluctant to examine.[428] As Goodpaster acknowledges, we need "clarity about the *framework* of moral consideration as much as about the *application* of that framework".[429] Whilst it may seem intuitive to presume the existence of other agents, this presumption needs to be argued for to avoid the sceptic's charge that we are simply begging the question in favour of the existence of other agents. Even if we can draw conclusions about agency in circumstances where there is sufficient evidence to do so, an additional question remains regarding which way to err when the evidence for agency is less conclusive.

---

[427] Known interchangeably as 'moral standing'. For a lucid discussion of existing accounts of moral status, see Warren (1997, Chapters 2–6).

[428] This is especially true within the tradition of Kantian and post-Kantian ethics. The possibility of other agents receives sustained attention in work of Levinas (1981) and Lacan (2006), who conceive an inter-relationship of self and 'Other'. These approaches represent a more recent post-modern current in which the experience of self implicates the existence of others. In this sense, one gains self-knowledge in part through experience of other agents. Insofar as the separateness of agents dissolves, such approaches have significant implications for ontology. However, they still do not provide an explanation as to why we are *morally* required to accept that there are other entities – human or otherwise – worthy of moral consideration.

[429] Goodpaster (1978: 309).

P. Bielby, *Competence and Vulnerability in Biomedical Research,*

Two issues here require our attention. First is the epistemological ground of agency ascription under the PGC. The second is how far this extends. Both concern how we come to identify agency-relevant characteristics and the moral status of those whose agency competence appears to be intermittent or partial. To explain each will allow us to identify when cognitively vulnerable human beings appear to be agents and whether we can treat them as potentially capable of developing task and decisional competences. More specifically, it will allow us to go on to consider whether cognitively vulnerable human beings have the potential for developing the competence to consent to biomedical research.

The aim of the present chapter is twofold – to explain the epistemology of agency ascription under the PGC and from that to draw some preliminary conclusions about which entities may be able to develop task and decision-making competences. My argument relies upon and develops the framework devised by Beyleveld, Brownsword and Pattinson.[430] The chapter begins by explaining the justification for accepting that there other agents apart from ourselves under the PGC, and then proceeds to examine whether entities who do not consistently display agency competence may nonetheless have the potential for developing specialised decisional competences.

The Principle of Proportionality[431]

In *Reason and Morality*, Gewirth claims that the distribution of the generic rights of freedom and well-being are not limited to those that have 'full' agency. Instead, the scope of the PGC permits proportionate allocation of the generic rights to partial agents[432] (or marginal agents[433]), including children, adults with intellectual disabilities and mental disorder, foetuses and non-human animals. The degree to which an entity can claim the generic rights follows from the degree to which the entity approaches being a 'full' agent.[434]

Gewirth expresses the Principle of Proportionality in the following terms. It is helpful if for 'Q' we substitute 'agency', and for 'R' 'the generic rights' under the PGC:

[430] Originally presented in Beyleveld and Pattinson (1998, 2000) and developed in Beyleveld and Brownsword (2001). References in this chapter will be to one or more of these instances in which the argument is presented.

[431] This section closely follows the structure of the discussion in Beyleveld and Pattinson (1998).

[432] The term Beyleveld and Pattinson (1998, 2000) use to describe those who do not possess capacities for voluntary action to the extent of full agents.

[433] The term used by James F. Hill (1984) to describe the same.

[434] *RM*: 121–128, 140–145.

> When some quality Q justifies having certain rights R, and the possession of Q varies in degree in the respect that is relevant to Q's justifying the having of R, the degree to which Q is had is proportional to or varies with the degree to which R is had. … Thus, if *x* units of Q justify that one have *x* units of R, then *y* units of Q justify that one have *y* units of R.[435]

Gewirth argues that this principle demonstrates the degree to which partial agents have the generic rights depends upon the degree to which they possess the generic capacities of action (the GCA).[436] The GCA are the practical abilities of the generic features of action which can be held by degree both above and below the level required for agency.[437] Gewirth also claims that proportionality underpins conventional accounts of distributive justice.[438] Whilst this latter claim is valid, the former argument is false. To explain this, let us explore in more detail a fundamental aspect of the PGC introduced in the last chapter.

The necessary and sufficient condition for possessing the generic rights is the ability to form a proactive evaluative relationship between oneself and one's own purposes. In this book, following Beyleveld, I will abbreviate this ability as 'rpvp'.[439] In order for any entity to have rpvp, the entity must possess the GCA to a level that would allow such a proactive evaluative relationship to exist. The condition of rpvp is absolute and does not vary.[440] Possession of the GCA above the level required for rpvp will bear upon task and decisional competences but makes no difference to that agent's moral status or the degree to which one is an agent – there are no 'super-agents' under the PGC.[441] Since only agents can hold the generic rights,[442] it follows that to have the GCA to the degree needed for rpvp is both a necessary and sufficient condition of having any generic rights *at all*.

As we saw in the last chapter, the PGC requires agents to claim rights because they are instrumental to the pursuit of their own ends or purposes, irrespective of what these may be. An agent need only value categorically instrumentally the generic needs of action, rather than agency as such.[443] It follows that the generic rights are a species of rights that correspond with the broader 'will theory' of rights (also known as the 'choice theory' of rights).[444] An immanent feature of agency under the PGC is that an agent must have the capacity to decide not to exercise one

---

435 *RM*: 121.

436 *RM*: 122, 141.

437 *RM*: 122.

438 *RM*: 121.

439 The abbreviation 'rpvp' denotes *relation: proactively valuing purposes*, which Beyleveld (1991) uses originally. I use it here for the sake of convenience.

440 *RM*: 123.

441 *RM*: 124–125, Beyleveld (1991: 250–256).

442 *RM*: 122–123.

443 *RM*: 136–137, 266, Beyleveld (1991: 28–30).

444 Discussed in Jones (1994: 32–36), Kramer et al. (1998: 195–232) and Edmundson (2004: 119–132).

or more of their generic rights or to waive the benefit of these rights.[445] Consequently, it is possible for an agent to waive the benefit of both a positive and a negative right that the PGC bestows,[446] such as the right to physical well-being (in refusing medical treatment) and, in certain contexts, the right to life (in suicide). To be able to waive the benefits of a right, one must have rpvp, the precondition of being a rights-holder under the PGC. However, partial agents by definition do not have rpvp. As having the GCA to the degree needed for rpvp is both a necessary and sufficient condition of having any generic rights at all, it is not possible to ascribe generic rights to partial agents.[447] Accordingly, Gewirth commits "the fallacy of disparateness"[448] in respect of his formulation of the Principle of Proportionality.

Beyleveld and Pattinson reject the Principle of Proportionality as formulated by Gewirth. They argue that it does not follow from x units of Q justifying that one have x units of R that y units of Q justify that one have y units of R, because:

(a) "either Q or R might be something that can be had entirely or not at all, but not in part;
(b) either Q or R assuming that both can be had to variable extents might not have a variable possession that is quantifiable in percentage terms, or even at all; or
(c) having 100% of Q, though sufficient to have 100% of R, might also be necessary to have any R at all."[449]

Beyleveld and Pattinson reconfigure the Principle of Proportionality as a necessarily true principle which states that proportionality between variable Q and R exists insofar as "the degree to which R is had *is a function* (emphasis added) of the degree to which Q is had".[450] There is no assumption of necessary proportionality between the two variables, but instead recognition of the logical inference between the two. This is the only interpretation of the Principle of Proportionality within the PGC that does not alter the ontology of the PGC.[451]

The fundamental practical difference between Gewirth's formulation of the Principle of Proportionality and the reformulation provided by Beyleveld and Pattinson is that Gewirth grants the generic rights to partial agents *qua* partial agents in proportion to their approach to full agency. Beyleveld and Pattinson, on the other hand, maintain that only agents as such have the generic rights, although

[445] *RM*: 334, Beyleveld (1991: 18).

[446] Beyleveld and Brownsword (2001: 72 and 74, n. 14).

[447] Beyleveld and Brownsword (2001: 118–119).

[448] The 'fallacy of disparateness' is expounded by Gewirth (1960: 313). One commits the error where one compares "fields or subject matters... on disparate levels or in disparate respects". Beyleveld and Pattinson (1998: 2).

[449] Beyleveld and Pattinson (1998: 9, 2000: 46).

[450] "When possessing some quality Q justifies having some property R, and it is *not the case* that the extent of having Q sufficient to justify R to the maximum extent possible is, at the same time, necessary to justify having R to any extent at all, the degree to which R is had is a function of the degree to which Q is had" Beyleveld and Pattinson (1998: 12).

[451] Beyleveld and Pattinson (1998: 49).

the PGC necessarily confers moral status upon apparent partial agents in proportion to their display of rpvp.[452] Whereas Gewirth argues for the moral status of partial agents, Beyleveld and Pattinson argue for the moral status of *apparent* partial agents.[453] This means that the moral status of any entity is proportional to the *probability* that it is an agent, not to the degree of approach to agency.[454]

However, this is not all which separates Gewirth's account of moral status from that of Beyleveld and Pattinson. In *Reason and Morality*, Gewirth appears to assume that we can know with certainty who are agents and who are partial agents. This rests upon a mistaken premise that if an entity behaves like an agent, it *actually is* an agent and if it does not behave in this way, it is not.[455] Beyleveld and Pattinson argue instead that our inability to access other minds dispels any such claims to certainty.[456] When applied under precaution, the revised Principle of Proportionality accounts for considerations that are involved in determining whether we should treat an entity as an agent.[457] Let us consider this more closely.

Precautionary Reasoning[458]

Agency itself is an ontological category, connected inextricably with an entity's ability to act. Although any agent can be certain of being an agent himself/herself/itself, our existential limitations and the problem of other minds means we can never be completely sure that another entity which behaves like an agent *actually is* an agent.[459] Indeed, it is impossible to prove beyond all doubt that any entity we take to be an agent is not simply an automaton, even though the entity's behaviour might seem intuitively to support an ascription of agency.[460] Behaviour and other empirical evidence that may be adduced for the existence of agency in some entity only seems relevant because it rests on a metaphysical assumption that there are agents other than myself.[461] Precautionary reasoning under the PGC provides a framework that sets out the circumstances in which we *should* treat some entity as

---

452 Beyleveld and Pattinson (1998: 3).

453 Deryck Beyleveld, personal communication, 19/12/01.

454 Beyleveld and Pattinson (1998: 29), Beyleveld and Brownsword (2001: 123).

455 Deryck Beyleveld, personal communication, 19/12/01.

456 Beyleveld and Pattinson (1998: 25–28). For influential discussions of the problem of other minds, see Ryle (1949), Malcolm (1958), Aune (1961) and Putnam (1979).

457 Beyleveld and Brownsword (2001: 119).

458 This section primarily draws upon ideas presented in Beyleveld and Pattinson (1998, 2000) and developed in Beyleveld and Brownsword (2001).

459 Beyleveld and Brownsword (2001: 120–121).

460 Beyleveld and Brownsword (2001: 120).

461 Beyleveld and Brownsword (2001: 120). From a neuroscientific perspective, the recent discovery of 'mirror neurons' in the human brain, which allow human agents (at least) the ability to share in the experiences of others, helps to explain this assumption. See Rizzolatti and Sinigaglia (2007) for a discussion.

an agent, given these conditions of uncertainty. The premise of the framework is the idea that the only type of claim we can make about another's agency is that it is possible or probable on the evidence we have available. Under precaution, the probability of agency increases with evidence for the characteristics of agency – this is the generic capacities of action (GCA) behaviour, which I will abbreviate to GCAb.

Ostensible Agents

Mindful of the above discussion, the best I can do is make an informed judgment that some other entity – human, animal, android or otherwise – appears, on the evidence, to be an agent. Such a judgment ascribes to the entity in question that it is *ostensibly* an agent. Ostensible agency is a metaphysical category rather than an ontological one, given the impossibility of adducing irrefutable empirical proof that there are other agents apart from myself.

Therefore, to the solipsist who doubts the existence of other agents, we can concede that it is no more possible to prove that some entity is an agent than to disprove it – even where the evidence suggests that the entity in question is an agent.[462] Nevertheless, the evidential proof for the proposition "this entity is an agent" has different moral implications from the evidential proof for the proposition "this entity is not an agent".[463] If I err in presuming the entity to be an agent, I restrict my exercise of the generic rights but I do not deny myself – or any other agent – the generic rights. However, if I err in presuming the entity not to be an agent, then I deny the entity the generic rights to which it is entitled as an agent.[464] Under the PGC, this is the gravest moral wrong that I can commit. The two scenarios respectively amount to a 'false positive' and a 'false negative' presumption of agency.

Although it is not dialectically necessary to hold that there are other agents, it is dialectically necessary for any agent to do as much as that agent can to avoid violating the PGC, where the avoidance of such violation is possible.[465] If my actions risk the possibility of violating the PGC – and I can avoid the possibility of this violation – assuming this risk itself violates the PGC.[466] Therefore, it is dialectically necessary for me to err on the side of presuming agency under these conditions of uncertainty, provided the consequences of this presumption measured in terms of the PGC are better than the consequences of not presuming agency.[467] Given these conditions of uncertainty, and the moral stakes involved in denying agency, the circumstances

---

[462] Beyleveld and Pattinson (1998: 26, 2000: 42).

[463] Beyleveld and Pattinson (1998: 26, 2000: 42).

[464] Beyleveld and Pattinson (1998: 25–26, 2000: 42). In doing so, I also incur self-contradiction as an agent through implicitly denying my own claim to the generic rights.

[465] Beyleveld and Pattinson (1998: 26, 2000: 42).

[466] Beyleveld and Pattinson (1998: 26, 2000: 42).

[467] Beyleveld and Pattinson (1998: 26, 2000: 42).

which would warrant a presumption against agency are very limited.[468] In other words, it is a general rule of the PGC that a false-positive presumption of agency is morally preferable to a false-negative presumption.

We can state the precautionary principle as follows:

> If there is no way of knowing whether or not X has property P, then, insofar as it is possible to do so, X must be assumed *to have* property P if the consequences (as measured by the PGC) of erring in presuming that X *does not have* P are worse than those of erring in presuming that X *has* P (and X must be assumed *to not have* P if the consequences of erring in presuming that X *has* P are worse than those of assuming that X *does not have* P.[469]

This entails the following categorically necessary principle, which stipulates the grounds on which we are logically committed to treating other entities as agents under precaution:

> Where X is an ostensible agent, the metaphysical possibility that X might not be an agent is to be wholly discounted, and X's ostensible agency is to be taken as sufficient evidence that X has the capacities needed to be an agent.[470]

This principle is, essentially, a moral argument for recognizing that there are other minds.[471] If this is accepted, then philosophical debate about the ontological possibility of other minds becomes, from a moral point of view at least, otiose.

Apparent Partial Agents

An apparent partial agent differs from an ostensible agent in the sense that it appears to have some of the generic capacities of action needed to at least some degree, but to an insufficient extent to display the ability to form a proactive evaluative relationship between oneself and one's own purposes (i.e. rpvp) at all.[472] Given an entity can possess the generic capacities of action at varying levels below that required for rpvp, the degree to which an entity displays the GCAb below that consistent with rpvp is proportional to the possibility that it is an ostensible agent.[473] Apparent partial agents may once have been agents, may become agents at some point eventually in the course of their existence or may never become agents, even though we cannot know this with certainty.

Such entities are *apparent* partial agents insofar as the evidence reduces the probability that they are agents. What is common to all apparent partial agents is

---

[468] The only justifiable circumstances I can imagine are where presuming agency in an entity would cause basic harm to other entities who are more probably agents, where the criterion of avoidance of more probable harm (see below) would apply.

[469] Beyleveld and Pattinson (1998: 27, 2000: 42–43).

[470] Beyleveld and Brownsword (2001: 121).

[471] Beyleveld and Pattinson (1998: 27, 2000: 42).

[472] Beyleveld and Pattinson (1998: 14).

[473] Beyleveld and Pattinson (1998: 50).

that, on the face of it, they appear not to be agents at all.[474] Of course, there remains a possibility, in proportion to the available evidence, that they could be agents, in spite of appearing incapable of displaying GCAb at a level consistent with having rpvp (GCAbrpvp).[475] That an apparent partial agent may once have been an agent or may become an agent at some point in the future contributes to the evidence base and raises the possibility.[476] At the outermost, one could construct a valid precautionary argument that plants, trees or bacteria *may* be agents whose biological composition or physical manifestation frustrates the display of GCAbrpvp (in spite of the counter-intuitive reaction this gives rise), although the probability of this being the case will be severely limited by the dearth of supporting evidence.

Ostensible Intermittent Agents

Ostensible intermittent agents differ from ostensible agents insofar as they are apparently partial agents some of the time and are ostensibly agents the rest of the time.[477] There are occasions on which the entity appears to display GCAbrpvp and other occasions when it does not. An ostensible intermittent agent is distinguishable from an apparent partial agent in the sense that an ostensible intermittent agent appears able to exercise rpvp some of the time and thus appears on the evidence to be an agent for as long as it can do this. An apparent partial agent appears unable to do this at any time and thus appears on the evidence not to be an agent at all.[478]

The frequency with which an ostensible intermittent agent displays rpvp determines the probability of it being an agent. Ostensible intermittent agents may once have consistently exhibited rpvp, may consistently exhibit rpvp at some point in the course of their existence, or may never consistently exhibit rpvp. Thus, another way of conceptualising the difference between an ostensible agent and an ostensible intermittent agent is that an ostensible agent appears to have rpvp as a consistent feature of its personality, whereas an ostensible intermittent agent does not.[479]

---

474 Beyleveld and Pattinson (1998: 14).

475 I use 'GCAbrpvp' in preference to Beyleveld and Pattinson's 'GCAbf' (which denotes the GCA behaviour in full) because GCAbrpvp more clearly denotes the required level of GCA behaviour for an ascription of agency. Given GCAb can be displayed at a level above that required for an ascription of agency, it follows that, strictly speaking, GCAbf corresponds to the highest possible level at which an entity could display GCAb (whether this could ever be known is a different matter), rather than merely to indicate the presence of rpvp.

476 I discuss the issue of potentiality within the section, 'The criterion of avoidance of more probable harm', below.

477 Beyleveld and Pattinson (1998: 15).

478 Beyleveld and Pattinson (1998: 15).

479 Beyleveld and Pattinson (1998: 15).

Evidence of Agency and Duties of Protection

How should an agent respond where an entity X exhibits GCAb at level lower than that which would suggest the presence of rpvp? On the evidence, X would be an apparent partial agent. Nevertheless, it is still not possible to infer that X is *not* an agent.[480] In the same way that I cannot have certain knowledge that X *is* an agent when X is an ostensible agent, equally, I cannot have certain knowledge that X is *not* an agent when X is apparently only a partial agent.[481] To illustrate the corollary of this, imagine I suppose that X is not an agent on the assumption that X is apparently only a partial agent. However, unbeknownst to me, X *actually is* an agent. In this case, I will have denied X the status of a rights holder under the PGC, and in doing so will have violated the PGC. Due to this persistent risk of misapprehension, it is *dialectically necessary* to do all one can to avoid this.[482]

When X is apparently only a partial agent, it is not possible for me to avoid the risk of denying X the benefits of the generic rights altogether. As Beyleveld and Pattinson argue:

> I can, indeed, refrain from harming (and can assist) X in ways that would safeguard the benefits that X would receive *if* X had the GR and chose to exercise them. I can, indeed, recognise duties not to harm (and to assist) X in various ways. However, it must not be forgotten that if X is, in fact, a PPA then the PGC requires X (thereby) to be accorded will claim-rights, the benefits of which X may waive. But, by not displaying the GCAbf [this use is equivalent to GCAbrpvp], X fails to demonstrate (even under precautionary reasoning) that X has the capacities by virtue of which it is able to waive the benefits of what it is entitled to. Thus, the "duties of protection" that I must recognise that I have towards X, where X is apparently only a partial PPA, are perforce paternalistic, which is at odds with what X is strictly entitled to *qua* being a PPA (should that be the case).[483]

Under precautionary reasoning, just as if X displays all the capacities of an agent this constitutes sufficient evidence that X is an agent, it follows that if X displays some but not all of the capacities for agency, then this constitutes insufficient evidence - but evidence all the same - that X is an agent.[484] Agents owe duties of protection to entities which are apparently only partial agents in proportion to the degree of approach to *ostensibly* being agents.[485] These duties are owed due to the *possibility* of them being agents, rather than them *actually* being partial agents.[486] That we owe duties of protection to entities that are apparently only partial agents entails that we must adopt a paternalistic approach as a substitute for their own capacity to act autonomously, as the entity would if it were ostensibly an agent.

---

480 Beyleveld and Pattinson (1998: 27, 2000: 43).

481 Beyleveld and Pattinson (1998: 27, 2000: 43).

482 Beyleveld and Pattinson (1998: 27, 2000: 43).

483 Beyleveld and Pattinson (1998: 27–28).

484 Beyleveld and Pattinson (1998: 29, 2000: 44).

485 Beyleveld and Pattinson (1998: 27, 2000: 45).

486 Beyleveld and Pattinson (1998: 33–34, 50, 2000: 45, 52), Beyleveld and Brownsword (2001: 123).

The position in respect of ostensible intermittent agents is more complex:

> Imagine X who is a partial PPA, developing towards having the quality rpvp as such. At some point in time, X develops all of the capacities needed to have the quality rpvp, but does so only ephemerally. At time t_1, X has the capacities for having rpvp. At a later time t_2, X does not display some of these capacities, and at a still later time t_3, X displays the capacities again. But X's state, as having the capacities for rpvp or not having them, varies in not altogether predictable ways. Gradually however, the frequency of the periods during which X has the capacities increases, until the time arrives when X has the capacities needed for having rpvp as a standing capacity of X's personality. During the time in which X is developing towards having rpvp as a standing capacity, X is an "intermittent PPA". When X has developed rpvp as a standing capacity, X has become a PPA-in-full in the sense of a full-time PPA.[487]

Here, a dilemma emerges. In the case of ostensible agents, precautionary reasoning enjoins that it is dialectically necessary to treat them as agents, unless they fail to display all of the capacities of action, in which case they are apparently only partial agents to whom we owe duties of protection. However, in the case of ostensible intermittent agents, it is dialectically necessary for us to act on the presumption that they are, for all practical purposes, agents.[488] This is because we have no reliable way of ascertaining the occasions upon which ostensible intermittent agents have the generic rights and when they have not.[489]

To illustrate, if we attempt to assess the agency status of David (who, unbeknownst to us, is actually an intermittent agent), we encounter two problems:

1. The test cannot be used to justify treating David according to the outcome of the test due to David's fluctuating agency status.
2. David's agency status determines the moral permissibility of the procedures used to assess David's agency status.[490]

For these two reasons, we will not be able to ascertain the status of David before any interaction with him. Since we cannot know at any point in time whether David is an agent or a partial agent, we must use precautionary reasoning. This stipulates that the negative consequences that follow from treating David as an agent when he is not are less than the negative consequences that follow from not treating David as an agent when he is. In the first case, if we assume that David is an agent, we will be attempting to impose duties on him that he is unable to discharge.[491] However, the obligations on agents that derive from the presumption that David is an agent do not deny the generic rights to others. Alternatively, if we presume David is not an agent when he is in fact an agent, then we violate the PGC as we deny David the protection of the generic rights.[492] Because a false positive

---

[487] Beyleveld and Pattinson (1998: 32).

[488] Beyleveld and Brownsword (2007a: 110).

[489] Beyleveld and Pattinson (1998: 32–33), Beyleveld and Brownsword (2007a: 109).

[490] Adapted from Beyleveld and Pattinson (1998: 32–33). See also Beyleveld and Brownsword (2007a: 109).

[491] Beyleveld and Pattinson (1998: 34), Beyleveld and Brownsword (2007a: 110).

[492] Beyleveld and Pattinson (1998: 34).

presumption of agency is morally preferable to a false negative presumption of agency, whenever an entity is ostensibly an intermittent agent, precautionary reasoning requires that we treat the entity as an agent.[493]

The Criterion of Avoidance of More Probable Harm

As stated, the PGC requires that we discharge our duties to respect the generic rights of all other agents. But since we can only draw inferences that any given entity is an agent from the evidence for agency it displays, how should we proceed if we are confronted by two entities, both of whom present similar evidence of GCAb and there is a particular good which cannot be distributed to them both equally?

Imagine that entity A is an adult chimpanzee and entity B is a two-year-old human being. Both are in need of an inoculation against a fatal virus, and only one inoculation remains.[494] Both display some evidence that they have occurrent purposes that they wish to pursue, although in neither case is there unambiguous evidence of the ability to form a proactive evaluative relationship between oneself and one's own purposes (rpvp) that a cognitively developed human being displays.[495] In the case of the chimpanzee, we know that its oblique display of the GCAb is the most it will ever be able to exhibit. In the case of the child, we know through our empirical awareness alone that it will in all probability display the GCAbrpvp as it grows older, which will almost certainly exceed the possibility of the chimpanzee displaying GCAbrpvp.

In such a situation, the PGC requires us to fulfil our duties towards those who are more probably agents in respect of singular or collective action to secure basic rights to well-being and freedom. It is only justifiable to exclude entities that are less probably agents from the benefits of PGC-protected rights where, in not doing so, we would compromise our ability to discharge our duties towards those who are more probably agents.[496] Although the PGC requires us to reach a decision on the probability of agency based on the evidence that is before us now, this does not mean that considerations of potentiality are entirely irrelevant, especially if the evidence indicates agency in a state of becoming.[497] In this example, we have reliable grounds to assume that the child has greater potentiality for agency through her

---

[493] Beyleveld and Pattinson (1998: 33–34), Beyleveld and Brownsword (2007a: 110).

[494] Let us suppose that this inoculation is as effective on apes as it is on human beings.

[495] For the purpose of this thought experiment, I will concentrate on the grounds for ascriptions of *intrinsic* moral status. This is in order to highlight how the criterion of avoidance of more probable harm treats entities *qua* the possibility of their being agents *simpliciter*. Thus, I will assume that neither the chimpanzee nor the child is the object of an ostensible agent's love or affection, as this could introduce problematic considerations of *derivative* moral status (generated by the significance which the ostensible agent attaches to the entity in question) into the analysis, which would obscure our appreciation of their inherent moral standing.

[496] Beyleveld and Pattinson (1998: 30–31).

[497] Beyleveld and Brownsword (2001: 125), Beyleveld and Pattinson (2000: 50–51).

nascent cognitive capacity, even though the evidence for agency she displays now is likely to be on a par with the chimpanzee. Conversely, the chimpanzee, appearing to lack the child's potential for ostensible agency, will not progress to a more advanced cognitive stage than that which he has now reached, and will remain apparently a partial agent for the rest of his life.

We can infer from this that the child is *now* more probably an agent, since what she will display eventually (ostensible agency) has its genesis in what she displays currently (apparent partial agency).[498] This is because the 'seed' of agency appears to be already evident in her cognitive functioning and, in the absence of profound intellectual disabilities, is likely to flourish over the coming years into full manifestation. From a precautionary perspective, we are required to treat the current stage of the child's development as frustrating her display of GCAbrpvp until a later stage of development. On this basis, the child's claim to the inoculation trumps the chimpanzee's claim – albeit narrowly – in moral terms. It is important to emphasise that this does not thereby negate the chimpanzee's entitlement to the inoculation. However, it does require us to give more weight to the claims of the child than those of the chimpanzee in resolving any competing claims.[499]

Beyleveld and Pattinson describe this principle as the criterion of avoidance of more probable harm:

> If my failing to observe a particular duty of protection to Z is more likely to mistakenly deny Z the status of an agent than is my failing to observe the duty of protection to X (and I cannot avoid failing to observe this duty to one of Z or X), then I ought to fail to observe my duty to X rather than to Z.[500]

Beyleveld and Pattinson frame the criterion in the context of duties of protection to X and Z (who are both apparent partial agents), but the criterion can extend to apply

[498] It is worth emphasizing that there is a difference here between a comparison of the moral status of the adult chimpanzee and the child and a comparison of the human embryo or foetus and the child. Were we to compare the status of the chimpanzee with the embryo or foetus, the status of the chimpanzee would evidently be greater. This is because of the minimal extent to which the 'seed' of agency appears to have developed in the embryo and foetus (especially in the early stages of gestation) and the remoteness to its realisation in childhood. By contrast, the adult chimpanzee almost certainly already displays the maximum level of GCAb it ever will, which, given the fallibility of our attempts at interpretation, could actually indicate possession of rpvp. The human embryo or foetus, as an entity physiologically dependent upon gestation in the womb, cannot ocurrently display GCAb, until at least the latter stages of pregnancy. For Gewirthian discussions of the moral status of the human embryo and foetus, see Steigleder (1998) and Beyleveld (2000).

[499] Beyleveld and Pattinson state this in the following terms: " .. if Y is apparently only a partial agent with y moral status (by virtue of Y's degree of the generic capacity for action) but not apparently a potential ostensible agent, and X is apparently a partial agent with y moral status and also apparently a potential ostensible agent, then agents must take more seriously the possibility that X is an agent than that Y is an agent, by virtue of which their duties of protection to X are greater than their similar duties to Y" (2000: 50–51).

[500] Beyleveld and Pattinson (1998: 29), Beyleveld and Brownsword (2001: 123).

to situations where X is an apparent partial agent and Z is an ostensible agent or an ostensible intermittent agent (or as we have seen above, a potential ostensible agent who is currently an apparent partial agent)[501].

The criterion of avoidance of more probable harm does not mitigate the conscientious effort we must make towards fulfilling duties of protection to as wide a range of possible agent-entities as is within our power. It would not absolve the moral failure of denying an entity the protection of the PGC protected rights where conscientious collective action could have meant that such a choice would never have arisen in the first place. In the above example, if the supply of inoculations had diminished to one dose due to waste, mismanagement or corruption, then the agents responsible for this state of affairs would also be morally responsible for the fact that the chimpanzee did not receive the inoculation. However, the agent responsible for making the choice between inoculating the child and the chimpanzee would not be morally responsible for this state of affairs (assuming that she had no part in these background factors) and would simply be acting in accordance with the PGC in less than ideal circumstances.

Degrees of Displaying GCAb and Its Relationship with Task and Decisional Competences

Ostensible Agents

As ostensible agents exhibit GCAbrpvp as a consistent feature of their personality, we have the strongest possible evidence that they agents. We acknowledged above that the degree to which agents display GCAb above the minimal level required for agency (i.e. for rpvp) will bear upon their task competence and decisional competences as agents.[502] Many ostensible agents will have the capability for developing an extensive range of task and decisional competences, which may take an overtly intellectual or practical form.

However, to claim a task or decisional competence right is not the same as to claim a generic right. A generic right – such as the right to life – is claimed merely as a consequence of being an agent, whereas a right to perform or participate in a specific task or make a specific decision can only be claimed though either demonstrable or presumed proficiency. As such, the latter variety of rights is contingent on specific features of one's agency, such as intellectual or physical ability or both. For a legal system to fail to recognise such a right where it is reasonable to presume proficiency or where demonstrable proficiency has been shown could violate

[501] Beyleveld and Pattinson (2000: 50–51).

[502] *RM*: 122–123, Beyleveld and Pattinson (1998: 13).

the agent's rights to additive well being under the PGC as it would needlessly frustrate the agent's level of purpose fulfilment.[503]

Legal systems typically bestow presumed task or decisional competence rights on all agents upon fulfilling a relevant criterion the agent can reasonably expect to achieve automatically (such reaching eighteen years of age in respect of the right to vote). Legal systems typically bestow demonstrable task or decisional competence rights upon an agent after reaching a manifest level of proficiency at that task (such as a licence to drive a car). In both cases, the task or decisional competence right is not bestowed by virtue of agency *per se* but by virtue of an 'agency plus' criterion. In the case of the right to vote, the right is bestowed due to automatically attaining an age where competence at that task or decision is presumed (i.e. agency plus presumed decisional competence to vote for a political party). In the case of the licence to drive a car, the right is bestowed as a consequence of having met a benchmark standard of skill at the activity (i.e. agency plus demonstrated vehicle operation competence and road use responsibility). With presumed task and decisional competences, an individual will have become an agent before reaching that age, and in many cases, already will have developed societal competence as well.[504] By contrast, in the case of demonstrable task or decisional competences, an individual will still be an agent (and retain societal competence, if she already possessed it), irrespective of whether she satisfies this level of proficiency, fails to demonstrate such proficiency, or chooses never to develop this proficiency at all.

Ostensible Intermittent Agents

As we saw earlier, ostensible intermittent agents are entities who are apparently partial agents some of the time and who are ostensibly agents the rest of the time. The threshold between an ostensible intermittent agent and an ostensible agent arises where the entity in question appears to display the generic capacities of action behaviour consistent with rpvp (GCAbrpvp) continually. The extent to which it is probable that an ostensible intermittent agent is an agent is proportionate to how frequently it displays rpvp.

Since any ostensible intermittent agent displays rpvp some of the time, they must be considered at least to have the developable potential for task and decisional competences, albeit at a lower level than an ostensible agent does. In practice, ostensible intermittent agents should be thought of as specific task or decision incompetent agents and in most cases will be treated as societally incompetent

[503] I use additive in the same 'exclusive' sense that Gewirth does in *RM*: 56–57.

[504] Indeed, societal competence (measured according to a substantively just standard) would be a prerequisite for the right to vote.

agents.[505] This means that we should treat ostensible intermittent agents as agents of restricted task and decisional competence who possess all the generic rights, but lack the abilities and independence necessary in order to prevent causing unintended harm to themselves or others. Returning to the earlier example of David, we should treat him as an agent yet ensure he receives support and assistance to act successfully as far as possible in relation to activities common to everyday societal interaction.

Insofar as the GCAb of an ostensible intermittent agent appears to fluctuate between GCAbrpvp and a lower level of GCAb, when the entity concerned evinces agency competence, its potential for developable task and decisional competences will be as high as it can ever be.[506] By definition, an ostensible intermittent agent will appear to have lapses of the GCA to a degree where it temporarily fails to appreciate the consequences of waiving the benefits of the generic rights to which it is entitled. However, not all of these lapses will be to the same degree. When considering the task and decisional competence of ostensible intermittent agents, we can draw a distinction between two types[507]:

Ostensible Intermittent Agents Who are Close to Having the Capacities Needed to Be an Agent in a Standing Fashion

This group is the closest that may approach full ostensible agency. Ostensible intermittent agents of this type consistently appear to understand the implications of waiving rights, but also experience periods where display of GCAbrpvp is questionable. As stated previously, precautionary reasoning denies that it is possible to determine exactly when the ostensible intermittent agent has the GCA necessary for rpvp, and when it has not. Since this type of ostensible intermittent agent is close to having the capacities needed to be an agent in a standing fashion, it follows that it would be extremely difficult to distinguish between occasions where it appears to have the task and decisional competences that it possesses as an agent and when it lacks them.

It would not be possible to re-evaluate the task or decisional competence of an ostensible intermittent agent who displayed rpvp almost all of the time, even if the individual actually did *not possess* rpvp at that point. As with an assessment of agency competence, by the time we had made a thorough assessment of an ostensible intermittent agent's specific task or decisional competence, it may well have returned to a state of possession of rpvp and ability to exercise those competences. To avoid denying such entities the protection of the PGC and infringing their rights to additive well-being, we should always treat such individuals as agents (albeit

[505] Beyleveld and Pattinson (1998: 47–49), Beyleveld and Brownsword (2001: 129).

[506] At least for as long as it continues to be ostensibly an intermittent agent.

[507] Beyleveld and Pattinson do not draw this distinction within their definition of ostensible intermittent agents.

with diminished societal competence) and respect the task and decisional competences that they display.[508] Allowing for natural variations in abilities, talents and skills, there would not necessarily be a considerable impairment in the range of task and decisional competences this type of ostensible intermittent agent could develop compared with ostensible agents.

Ostensible Intermittent Agents Who are Less Close to Having the Capacities Needed to Be an Agent in a Standing Fashion

This second type of ostensible intermittent agent frequently appears to understand the implications of waiving rights but experiences protracted phases where it appears not to be able to exhibit GCAbrpvp. In other words, it is quite plain to observe when this type of ostensible intermittent agent is ostensibly not an agent and, by extension, unable to exercise task or decisional competences.

It follows that there will be times during which this type of ostensible intermittent agent not only fails to display rpvp but also probably fails to possess rpvp.[509] During these times, it will be unable to exercise intentional control over its actions and will sometimes behave in ways that cause unintended harm to itself or others because of its apparent inability to act freely and voluntarily. This means that ostensible intermittent agents who are less close to having the capacities needed to be an agent in a standing fashion will have a reduced potential for developing and exercising the abilities to undertake specific tasks and make specific decisions, precisely because the periods of time in which they cannot exercise these abilities are more extensive.

Under precaution, we should treat as agents all ostensible intermittent agents who are less close to having the capacities needed to be an agent in a standing fashion. However, more compelling grounds exist for restricting the range of tasks and decisions that such ostensible intermittent agents are permitted to undertake than with those who are closer to having the capacities needed to be an agent in a standing fashion. Ostensible intermittent agents who are less close will have an impaired ability to develop task and decisional competences that require consistency in behaviour or sustained concentration over time.[510] They would be more prone to causing unintended harm to themselves and others if we were to permit them to develop or exercise these competences. Accordingly, their societal incompetence will be more visible, and we would be justified in treating them as societally incompetent agents.

[508] However, it may be necessary to monitor the exercise of these competences more closely for the sake of that ostensible intermittent agent *qua* agent and for the sake of other agents.

[509] Although of course, we cannot know this with certainty.

[510] Such as cooking a meal, driving a car or reading a novel.

Apparent Partial Agents

As we have seen, the extent to which an entity that is apparently only a partial agent exhibits GCAb determines the possibility that it is an agent. However, at no point do apparent partial agents exhibit behaviour that is commensurate with having rpvp, which is required to be able to develop or exercise any kind of task or decisional competence. Even if an apparent partial agent is actually an agent without the capacity to display rpvp (remembering that we cannot ultimately prove or disprove this), we could not treat an entity without this capacity as having or being able to develop task or decisional competences. The risks attached to doing otherwise, measured in terms of the harmful consequences for the generic rights of entities who are more probably agents, and for the generic rights of the entity in question *if* (unbeknownst to us) it is an agent, would simply be too great. For these reasons, we cannot treat apparent partial agents as the bearers of task or decisional competences.

Summary

Gewirth's Principle of Proportionality cannot claim for itself the status of a necessary truth. On this basis, we cannot use it to delineate the Gewirthian moral community. Alternatively, Beyleveld and Pattinson's reformulation of the Principle of Proportionality as a necessarily true principle, combined with the dialectical necessity of precautionary reasoning, offers a compelling response to the central themes of this chapter. First, it offers a dialectically necessary argument for making ascriptions of agency competence. Second, it repudiates the claim that we are begging the question in favour of the existence of other agents – to suppose the existence of other agents is a moral requirement even if the proposition is ultimately incapable of metaphysical proof. Third, we can apply the criterion of avoidance of more probable harm to resolve conflicts between discharging duties of protection to entities that are more probably and less probably agents.

The categories of ostensible agency, ostensible intermittent agency and apparent partial agency are ascriptions of the possibility and probability of agency rather than metaphysically verifiable statements about the agentive properties of that entity. The duties that we must discharge to an entity falling under any of these categories are predicated on the possibility of it being an agent rather than it actually being a partial agent (as Gewirth would have it). As we saw in the example featuring the child and the chimpanzee, the duties to entities that are more probably agents supersede duties to entities which are less probably agents where it is not possible to discharge both sets of duties. This grounds an account of moral status under precaution.[511] When in doubt as to whether we should treat some entity as an

[511] Beyleveld and Pattinson (2000: 44).

agent, we should resolve this dilemma in favour of presuming agency, unless to do so would risk a violation of the generic rights of another entity or entities that are more probably agents. However, this latter option must only be pursued where it is supported by reliable and comprehensive evidence to justify the relative status of the entities in question and to explain which of the more probable agent's generic rights would be violated if the presumption were applied. Even if we err in making an ascription of agency, the consequences of a false positive ascription are usually better (measured in terms of the PGC) than a false negative ascription.

From this, we can reframe the principle established in Chapter 1 that only agents can be the bearers of task and decisional competences in terms consistent with the PGC. We should treat ostensible agents and ostensible intermittent agents under precaution as having at least the potential to develop task and decisional competences due to their consistent or partial display of the ability to form a proactive evaluative relationship between oneself and one's own purposes (rpvp). We cannot treat apparent partial agents in the same way due to the complete absence of display of rpvp, although we still owe important duties of protection to them as possible agents in proportion to the extent to which they approach ostensible agency.[512] The extent to which an agent displays the generic capacities of action behaviour (GCAb) above the level required to exhibit rpvp (i.e. above GCAbrpvp) will determine the range of task and decisional competences that agent will be able to develop. This applies both to competences that are presumed legally (on the agent's satisfying a particular criterion) and to those that require demonstration before legal recognition is granted.

To deny an agent the right to develop or exercise a task or decisional competence may undermine the agent's additive rights to well-being, although the negative consequences of this are eclipsed by the negative consequences that follow from mistakenly denying the entity the status of an agent altogether. There are also other considerations, such as the possibilities of causing unintended harm to self and others which will place reasonable limits on the contexts in which such competences are officially recognised. In this chapter, we have considered these issues primarily in the abstract. In the next chapter, we will open this analysis out to consider how we should treat actual groups of cognitively vulnerable human beings under precaution, in order to begin the process of considering the decisional competence of such cognitively vulnerable individuals to consent to biomedical research.

[512] Beyleveld and Pattinson (1998: 29), Beyleveld and Pattinson (2000: 45).

Chapter 5
The Competences of Cognitively Vulnerable Groups

Making an informed judgment about the competences of any group of possible agents involves two elements. First, we need to elaborate an ethical justification of the possibility of these competences. Such justification requires us to decide whether we are right to presume certain kinds of competences or to allow the development or exercise of a particular competence amongst a group of individuals. The measure of ethical justification is a precautionary application of the PGC. The second element is an empirical investigation into the possibility of members of that group developing or exhibiting a particular competence. The central task here is to explain how the physiological and psychological position of the possible agent may influence their agency, task and decisional competences. Together, this allows for a 'reflective equilibrium' between what we should assume about members of the group on ethical grounds and what capabilities empirical research has shown they possess. If there is a tension between the two, then we should usually consider revising our ethical understanding to take account of the empirical evidence, unless we have good reason not to (for example, where the empirical evidence is unreliable).

This process is especially important for cognitively vulnerable individuals. Judging the presence or absence of agency, task or decisional competences may be harder with cognitively vulnerable individuals than with individuals who lack this form of heightened vulnerability. Alertness to fluctuations in ability, unusual ways of expressing competence and findings from psychiatric studies will bear upon our judgment here. However, for such judgments to have integrity, we cannot treat cognitively vulnerable individuals as one amorphous group. To do so would fail to draw out the way in which specific cognitive vulnerabilities such as depression or schizophrenia interact with the display of agency competence and with the development and exhibition of task and decisional competences. It also would not leave us any clearer about which cognitively vulnerable individuals are likely to have the occurrent or developable competence to consent to biomedical research, and which have not.

In this chapter, I apply the precautionary reasoning framework set out in the last to consider the agency, task and decisional competences of five different groups of cognitively vulnerable individuals. In particular, I consider the relationship between agency, task and decisional competence and cognitive vulnerability by reference to psychiatric nosologies and debates in developmental psychology, ethics and the philosophy of mind. The aim is to offer an account of the relationship between cognitive vulnerability and the likely existence of these competences. As I stated in the introductory chapter, the five groups of cognitively vulnerable human beings on which I will focus in this book are older children and adolescents, adults with intellectual disabilities, adults with depression, adults with schizophrenia and adults with Alzheimer's disease and other dementias. The choice of groups achieves a balance between examining a diversity of different cognitively vulnerable groups and maintaining a clear focus to allow for detailed analysis of each within the structure of the book.

By the end of the chapter, we will be able to reach a position on the likelihood of occurrent competences and potential for prospective competences we can justifiably ascribe to these five groups of cognitively vulnerable individuals from a precautionary perspective. This position will inform the discussion in the next chapter, where I consider whether some, if any, members of these cognitively vulnerable groups may have at least the potential for developing decisional competence to consent to medical research, according to psychological and psychiatric studies.

Older Children and Adolescents

In this book, I take the category 'older children and adolescents' to refer to young people between the ages of around twelve to sixteen rather than to denote a legal category. By extension, I will use the term 'adulthood' in a purely developmental sense to refer to a state of maturation commonly reached by sixteen years of age where a young person becomes societally competent. This is to avoid the contingencies of legal categories of adulthood, as the threshold age at which the attainment of legal adulthood (or legal majority) is reached can vary.[513] Accordingly, I do not intend it to denote the age of legal adulthood in any jurisdiction. Before we consider the competences of older children and adolescents, it is important to draw some salient distinctions between neonates, infants, older children and adolescents in order to be clear about how precautionary reasoning applies.

Gewirth views children in respect of their proximity to full agency as "potential agents, in that, with normal maturation, they will attain the characteristics of control, choice, knowledge, and reflective intention that enter into the generic features of action".[514] In relation to neonates and infants (aged from birth to around four years of age), Gewirth is right insofar as they await the cognitive, emotional and physiological developments that will transform their potential for agency into full ostensible agency. Because the capacity to form and pursue purposes emerges gradually over a number of years rather than all at once, neonates almost certainly do not have this capacity, whilst infants are subject to cognitive limitations such as an under-developed awareness of the future and an inability to identify values and

[513] Clucas (2004: 87–88 and 89) also argues that societal competence is a more preferable approach to understanding the significance of adulthood than legal majority.

[514] *RM*: 141.

manipulate abstract concepts that bear on the process of action.[515] It follows that younger infants will display degrees of the generic capacities of action behaviour (GCAb) somewhat below the level required to exhibit the ability to form a proactive evaluative relationship between oneself and one's own purposes (GCAbrpvp), whereas older infants will display degrees of GCAb closer to GCAbrpvp and exceptionally, GCAbrpvp itself. Since both will depend upon cognitively mature agents, such as parents, carers or teachers for their guidance and welfare, it is difficult to claim that infants *adopt* the purposes of others because they are in most cases apparently incapable of the voluntariness and freedom that would allow such adoption.[516] On this basis, we should treat neonates and most infants as apparent partial agents who are potential ostensible agents under the PGC.[517]

Gewirth is also right that children gain the GCA as they develop. However, recent work in developmental psychology and the philosophy of mind challenges Gewirth's association of partial agency with childhood.[518] Even on conservative estimates, most children will have reached full ostensible agency between seven and eleven years of age and it is conceivable that children could reach ostensible agency between the ages of three and five.[519] That children pursue freely chosen purposes in their learning, play and interaction with family, peer group and teachers provides strong support for this claim. Thereafter, the level of GCA above that required for rpvp will increase and become progressively more sophisticated as the child grows older. By the time that cognitively functional children reach adolescence and the cusp of adulthood, they will consistently display GCAbrpvp, and frequently exhibit GCAb to a degree that exceeds the minimal level required for agency.[520] In adolescents, capacities for action may be in some cases more sophisticated than in adults.[521] This is manifest in the older child and adolescent's ability to cultivate and exercise specialised task and decisional competences, and ultimately fully-fledged societal competence.

As there is no definite point at which a child begins to display GCAbrpvp, there will be significant variance between the agentive abilities of different children. On the face of it, in every case where a child displays rpvp, precautionary reasoning

---

[515] For a discussion of these issues, see Keenan (2001, Chapters 6–9).

[516] An agent by definition must be able to value proactively the purposes that she has chosen, even if she has adopted these purposes from others.

[517] I extrapolate this from the analysis in Beyleveld and Pattinson (2000, esp. pp. 50–51). Clucas (2004) argues that neonates and young children are 'practical agents' and 'immature agents' respectively. I agree with Clucas that Gewirth and possibly even Beyleveld, Pattinson and Brownsword pitch too highly the level at which ostensible agency is reached.

[518] For a detailed analysis, see Alderson (1990, esp. Chapter 5), Wellman (1992), Astington (1993), Mitchell (1997) and Clucas (2004).

[519] Beyleveld and Brownsword (2001: 129). The seminal work in this area is Piaget (1950, repr. 2001), which influenced Kohlberg (1984) and led to Gilligan's feminist critique (1982). Although Piaget's theory has been widely criticized, it nonetheless remains influential.

[520] Cf. Piaget's 'formal operational stage' and Vygotsky's 'higher mental functions'. For a discussion, see Keenan (2002: 129–133).

[521] Weithorn and Campbell (1982).

requires an acknowledgement of their ostensible agency. The PGC's dynamic-instrumental justification of legal protection[522] also demands official recognition of their task and decisional competences and when appropriate, their societal competence. To fail socially or legally to recognise task and decisional competences is a *prima facie* denial of their generic rights to additive well-being. If we are not to risk this denial, then it is incumbent upon us to create laws and to design institutions that respect all the agency rights of members of this age group, including rights to additive well-being.[523] Nonetheless, the importance of this should not be viewed in isolation from other societal aims. To pursue a policy of individual recognition of young people's task and decisional competences could lead to more serious implications (as measured by the PGC) than not recognising these rights.

To appreciate this, let us first suppose that we introduce a policy of according legal rights of specific task competence, decisional competence or societal competence on a case-by-case basis to any young person who displays GCAb above the minimal level required for agency. This would require each young person to undergo detailed assessment to ascertain whether she possesses the competence in question and could exercise it to the satisfaction of the assessors. The rigor of the assessment procedure would increase with the complexity of competence in question. This would inevitably involve a significant use of administrative time and resources.

Within a PGC-compliant polity, administrative burdens do not constitute sufficient reason for not implementing a policy, particularly if the official recognition of an agent's societal competence is at stake.[524] However, this does not imply that administrative burdens involved in respecting the rights of some agents cannot have repercussions that are more serious for the rights of other agents under the PGC.[525] It is highly probable that deployment of such resources for this purpose would interfere with the generic rights of other ostensible agents – adults and children – to have certain types of basic and non-subtractive goods provided by the state.[526] This would be so even if, as a PGC-compliant polity would require, tax rates were high and levied progressively.[527] If a PGC-compliant legislature had to choose between funding a case-by-case competence recognition programme and another to provide sheltered accommodation for the homeless, to fund drugs to treat breast cancer or to sustain community support programmes for people with mental disorder, the deployment of resources should be allocated to fund one of the latter objectives,

---

[522] Gewirth (1982: 61–62). See Chapter 3 for discussion.

[523] This is perfectly compatible with the claims that children and adolescents require support for their emerging autonomy and protection from exploitation. See Clucas (2004: 91–95).

[524] Clucas (2004: 92).

[525] For a wider discussion of the regulatory problems encountered in a PGC-compliant polity, see Beyleveld and Brownsword (2006).

[526] Such as education, housing, welfare, etc.

[527] *RM*: 315–316, 325. For a discussion of the incompatibility of the PGC with modern capitalism, see Beyleveld and Brownsword (1986: 307 and 452).

rather than the first.[528] We should only pursue case-by-case competence assessment programmes where the effects of such a policy do not threaten directly or indirectly the violation of more important PGC-protected rights.

In any event, case-by-case competence assessment becomes less important in terms of adolescents and young people on the cusp of adulthood, where threshold tests of task, decisional or societal competence may be used. The PGC requires that policy makers and legislators set these thresholds after careful consideration of the consequences for the generic rights of all agents affected. Once the majority of that group begin to display competence to decide in those spheres, any PGC-compliant legal system ought to give effect to those rights.[529] This argument supports the position of those who are keen to reduce the age at which the state politically enfranchises young people from eighteen to sixteen. It also causes us to reconsider the value of paternalistic practices in other areas of public life, such as the fetters placed on sixteen and seventeen year olds refusing life-sustaining medical treatment in England and Wales.[530]

Adults with Intellectual Disabilities[531]

By 'adults with intellectual disabilities', I refer to a range of mentally impairments that fall within the definition of mental retardation offered by the World Health Organisation ICD-10 nosology. This states:

> Mental retardation is a condition of arrested or incomplete development of the mind, which is especially characterised by impairment of skills manifested during the developmental period, which contribute to the overall level of intelligence, i.e. cognitive, language, motor and social abilities.[532]

---

[528] The likelihood that, in early adolescence at least, few children would be found to be societally competent strengthens the grounds for this decision.

[529] Or even one committed to human rights. See Beyleveld (1996).

[530] Discussed in Chapter 7.

[531] For reasons of analytical clarity and out of respect for the different experiences and identities of each group, I distinguish adults with intellectual disabilities from adults with forms of mental disorder. Moral and political philosophers do not universally adopt this approach, however. Some bracket intellectual disabilities together with mental illness when discussing agency and decisional competence, and use anachronistic nomenclature. Both of these shortcomings are evident in Wellman's analysis of "the mentally limited" (1995: 126–132). Lomasky (1987: 202–212) emphasises the need to distinguish those who have lost the ability to pursue freely chosen ends from those who never had the capacity, but places both within the awkward epithet "defective human beings" (*ibid.*: 202). Gewirth unhelpfully refers to "the insane and other such mentally deficient persons" when discussing the agency status of adults with mental impairment and mental health problems (*RM*: 120), which Cavadino has rightly described as "insensitive" (1997: 240). Viewed together, this is scarcely an improvement on Locke's "idiots and madmen" (1997: 156–157) three centuries before. I use the term 'adults with intellectual disabilities' to avoid any pejorative connotations.

[532] World Health Organisation (1992: 226).

Mental retardation, or intellectual disability, is not an absolute but a scalar quality. Impairment may be concentrated in specific areas of mental activity and need not be global or experienced to the same degree. There are a number of permutations of this. For example, one individual may suffer from cognitive impairment yet possess normal motor or social capabilities, another may experience cognitive impairment to a lesser degree with significantly impaired motor abilities but have no impairment of social capabilities. In the case of adults with intellectual disabilities, it is helpful to be clear about what kind of intellectual disability would still give rise to a display of GCAbrpvp, and when disabilities of this type may prevent its display.

In the fourth edition of the American Psychiatric Association's *Diagnostic and Statistical Manual of Mental Disorders* (DSM-IV), mild mental retardation corresponds to an intelligence quotient (IQ) level of approximately 50 or 55–70; moderate retardation from 35 or 40 to 50 or 55; and severe mental retardation from 20 or 25 to 35 or 40. Profound mental retardation is an IQ falling below 20 or 25.[533] Individuals with an IQ of less than 40 will have considerable difficulties comprehending and engaging in basic human interaction. Higher intelligence quotients, particularly between 60 and 70, will predispose an individual to greater levels of comprehension and understanding, which education and therapy can improve.

Despite the widespread measurement of intellectual (dis)abilities in terms of IQ, we should treat this approach with scepticism. IQ is unreliable as a measure of accounting for the myriad ways in which each human being can demonstrate their intelligence, as it is narrow, class-biased and culturally specific.[534] Although basic forms of intelligence that comprise voluntariness and purposivity will be required for agency, the ability to act for freely chosen purposes is not necessarily contingent on IQ.[535] Ascriptions of agency to adults with intellectual disabilities turn on whether they appear, under precaution, capable of acting *at all*, not whether they are capable of doing so according to a socially constructed standard. If we remember that the ability to act requires, at its most fundamental level, nothing more than to value an end and to value instrumentally the means to that end, then it is difficult to see how this could be measured solely by a test of linguistic, arithmetical and logical reasoning. An adult with an intellectual disability who, for example, attends a day centre to meet other adults in a similar position to herself or pursues her favourite hobby of painting acts for freely chosen purposes, irrespective of where her IQ happens to lie on the scale. Although we cannot discount the implications of IQ measurement entirely – an IQ falling below 20 or 25 could inhibit the display of GCAbrpvp – in this book, I will take degrees of intellectual disability to refer to degrees of display of GCAb rather than IQ.

Under precautionary reasoning, there would be little difficulty involved in making an ascription of ostensible agency to individuals with mild intellectual disabilities. The minor diminution in cognitive capacity will leave their display of GCAbrpvp

---

[533] American Psychiatric Association (1994: 40).

[534] See for a general critical discussion, Block and Dworkin (1977).

[535] Although specific task and decisional competences may be connected with intelligence.

unaffected. Impairment of GCAb above the level required for rpvp may affect the development of certain task or decisional competences but would be unlikely to affect societal competence. Alternatively, individuals who experience moderate intellectual disabilities are still ostensibly agents, but compared to individuals with mild intellectual disabilities, their range of task and decisional competences will be more heavily impaired. This is likely to affect their societal competence.

We should treat individuals who experience severe or profound intellectual disabilities as ostensible intermittent agents who are close to having the capacities needed to be an agent in a standing fashion, or ostensible intermittent agents who are less close to having the capacities needed to be an agent in a standing fashion. This depends upon how the intellectual disability concerned manifests itself and the amount of time during which they do not display GCAbrpvp. They would be almost certainly societally incompetent and would be significantly limited in their display of task and decisional competences and their apparent ability to develop them. Viewed together, the response of precautionary reasoning towards adults with intellectual disabilities affirms the expectation that the probability of such an individual displaying societal competence will vary in inverse proportion to the level of intellectual disability. It also strongly supports the widely held contemporary view that many adults with intellectual disabilities are, either alone or with appropriate support, capable of making decisions for themselves.[536]

It is uncontroversial that we should treat individuals experiencing social forms of intellectual disability as agents under precautionary reasoning. Most adults who experience these forms of intellectual disability, such as autistic spectrum disorders (ASDs) including Asperger's syndrome, are entirely unaffected in respect of their abilities to act for freely chosen purposes.[537] It is also highly likely that they could develop, for example, academic competence to an equal degree as an agent who does not experience such disabilities, in the same way that an agent with language impairment could still become an accomplished cellist. We should only consider an inability to cope with social interaction as impeding societal competence if it leads an individual to cause unintended harm to herself or others. In these cases, we would be right to judge an individual who experiences a severe social form of intellectual disability to be in greater need of assistance in everyday interpersonal situations and to have a more limited range of actual or developable task or decisional competences. Such a condition may actually indicate a deeper level of intellectual disability,[538] in which case the response to this under precautionary reasoning should be the same as that towards other individuals with intellectual disabilities.

The issue of adults who are likely to possess rpvp but whose means to display this is limited through motor impairment warrants particular consideration. Cases

[536] See Cantor (2005: 28). Gunn et al. (1999) provide a significant multi-disciplinary investigation and analysis which affirms this view.

[537] Frith (2004). For discussions that evaluate the ASD sufferer's range of skills and abilities, see Hill and Frith (2004) and Gillberg (2002).

[538] Frith, *ibid.*

such as motor neurone disease epitomize this, where the motor retardation of sufferers prevents them from enunciating speech or writing text, even if they are perfectly capable of comprehending language and ideas.[539] In this context, let us consider Jocelyn, a fictional motor neurone sufferer in the advanced stages of the disease. She is perfectly capable of comprehending language and ideas, and manipulating abstract concepts, although there is no means for her to communicate other than using a keypad-controlled voice synthesiser. Her intentions, decisions, beliefs, reasons and emotions have only the conduit of modern technology through which to express themselves.

That we should treat Jocelyn as an ostensible agent under the PGC is beyond doubt. It seems clear that she has rpvp and is able to act to achieve freely chosen ends. Since in any case we cannot have direct access to her mental state, all the evidence can do is to build a more complete picture that she is, on the balance of probabilities, an agent. When building this picture, typically our focus will be on how well Jocelyn can express her agency competence in optimum conditions (be it through a voice synthesiser, sign language or any other means). The knowledge of how Jocelyn's behaviour would be frustrated in the absence of the artificial means of display contextualises our empirical observation of her behaviour when given those means of display. Jocelyn's behaviour with the voice synthesiser, then, is clearly dispositive in erring on the side of presuming agency.

Nevertheless, we can go further than this. Where we have reliable grounds for believing that a human being or any entity retains an inherent capacity for agency, but lacks the means to express this independently, precautionary reasoning requires we should continue to treat that entity as an agent.[540] In making this judgment, we should draw on as much information as is available about the entity's mental state, in past, current and future periods.[541] We are morally enjoined to weigh such evidence for agency more strongly than evidence against agency, even where we can present evidence to show that the entity may temporarily lack rpvp at a particular point in time and that it is an ostensible intermittent agent (such as a human being

[539] The theoretical physicist Stephen Hawking and 'right-to-die' litigant Diane Pretty are good examples of this.

[540] This is analogous to the reasons for presuming that my partner is an agent when she is asleep. She retains her *prospective* ability to act even though she is not able to act now due to an altered state of consciousness. The overwhelming evidence that she is an agent whilst awake and the knowledge that her display of rpvp will return on waking will always override any countervailing claim that she is not an agent whilst she sleeps because she ocurrently does not display rpvp (when the effects of acting on both propositions are measured by the PGC). See also the discussion in Beyleveld and Brownsword (2007a: 105–106).

[541] This raises issues of psychological connectedness and continuity within debates surrounding personal identity and the 'unity' of selfhood. Even if we were to concede Parfit's argument (1984: 302–307) that my future self essentially represents a different person from that which I am now, we would still have to address the question of whether this has any bearing on agency, since agency does not necessarily presuppose a 'self' in Gewirthian theory. This is particularly so if we take the 'self' to mean "an embodied repository of integrated psychological states" (Radden, 1996: 11), which is particular to human beings, rather than merely a "mental presence" (Strawson, 1997: 407).

in a catatonic state).[542] It follows that we must do all we can to act on the presumption that the entity is an agent, unless we have a conflict between discharging our duties to an entity that is more probably an agent, in which case the criterion of avoidance of more probable harm applies.

To illustrate this, consider the example of a human coma sufferer. She has the means to display agency independently but has apparently lost agency competence, at least in the short term. We ascertain the agency status of the coma sufferer given the neurological evidence that the coma sufferer probably lacks the capacity for intentionality[543] and with knowledge of how the coma sufferer would have been able to use her intact means of display (capacity for gestures, voice projection, etc) had she not have lapsed into a coma.[544] This leads us to the conclusion that, for the time being at least, the coma sufferer is apparently only a partial agent.

When applied to Jocelyn (and others like her), neither approach for ascertaining agency status is relevant. Even where she lacks the technological aid, she can still raise a finger to communicate.[545] This means that there is no neurological evidence to suggest she lacks rpvp when unassisted.[546] Nor is it possible to suggest that the abilities of display are intact but dormant, as the nature of her condition means they have been lost irrevocably. Instead, we ascertain Jocelyn's agency status through an appreciation of her *latent* rpvp. This requires us to broaden the principle elaborated earlier, that the probability of rpvp possession informs ascriptions of agency. In most cases, *display* of GCAbrpvp determines this. However, in cases such as Jocelyn's, it is enough that we have knowledge of her enduring capacity for autonomous action.[547] From the perspective

---

[542] That it is possible (even if not probable) that such an entity will regain rpvp at a later date is evidence which should incline us to treating that entity as an agent, even though the moral weight attached to the possibility of regaining rpvp will in most case be stronger than the scientific likelihood of it happening.

[543] Although a recent study with a patient in a vegetative state lead by Adrian Owen of the Cognition and Brain Sciences Unit at The University of Cambridge found evidence of brain activity in response to suggestions to imagine engaging in physical activity, such as hitting a tennis ball. The authors note "her [the patient] decision to cooperate with the authors by imagining particular tasks when asked to do so represents a clear act of intention, which confirmed beyond any doubt that she was consciously aware of herself and her surroundings" (2006: 1402). Even if subsequent research were to cast doubt on these findings, the moral weight that precautionary reasoning attaches to them is considerable.

[544] This counterfactual proposition assumes that the coma victim had not sustained physical injuries that would compromise the means of display.

[545] Even the most limited form of communication – the systematic blinking of an eyelid to indicate 'yes' and 'no' – still evinces rpvp under precautionary reasoning. This position is affirmed in English law in *Re AK (Medical Treatment: Consent)* [2001] 1 FLR 129.

[546] Although were she to become fully 'locked in', her cognitive capacity may then only be able to be ascertained using an EEG-based brain-computer interface, the success of which is currently unclear. See the study by Hinterberger et al. (2005).

[547] This knowledge would have to be obtained through rational investigation amenable to logical scrutiny – it could not be knowledge gained by superstition, prejudice or 'parroting'. Cf. Dworkin's criteria for moral positions elaborated in *Taking Rights Seriously* (1978, Chapter 10).

of precautionary reasoning, that she cannot sustain rpvp behaviour herself is of no consequence. It follows that we should treat Jocelyn as an agent just as much in circumstances where the electronic apparatus to facilitate her communication is absent.

However, insofar as assistance greatly enhances Jocelyn's communication capabilities, Jocelyn has a generic right under the PGC to assistance in communication. How is this so? Precautionary reasoning requires us to work on the proposition that Jocelyn is an agent, but whose heightened vulnerability arises from the contingent circumstances of not being able to communicate effectively without assistance. Under the PGC, communication amounts to the capacity for purpose fulfilment (a non-subtractive good). Moreover, successful communication may bear upon the having of mental equilibrium (a basic good).[548] It is also a prerequisite for being considered, as Beyleveld and Brownsword put it, as "an ideal-typical subject of consent".[549] To suggest that an agent may not claim rights to assistance with communication would constitute a failure to recognise why communication is valued under the PGC.

As we saw in Chapter 3, all agents can expect to receive support from others in securing the generic rights under the PGC. This applies in respect of the fulfilment of basic, non-subtractive and additive rights to well-being and freedom.[550] As a matter of PGC-governed social justice, other agents are under a duty to assist agents like Jocelyn in exploring effective methods of communication, in order to allow her to communicate with us and for us to communicate with her. This duty bears especially upon those individuals who have a particular responsibility for her welfare, such as carers or medical practitioners.[551] By extension, we can state that in circumstances where such communication aids are not available, agents owe duties of protection to Jocelyn to protect her from the harmful and exploitative consequences of not being able to communicate. This represents two general principles that apply in respect of the moral treatment of any entity with latent or heavily restricted means of display of rpvp.

Adults with Depression

There are several varieties of depression contained within psychiatric nosologies.[552] The DSM-IV definition of a 'major depressive episode' captures many of the common symptoms of depression, which I will take as the focal meaning for the purposes of this book:

---

[548] The connection between successful communication and the maintenance of mental equilibrium is, I think, too strong to be speculative.

[549] Beyleveld and Brownsword (2007a: 100–101).

[550] *RM*: 67, Gewirth (1996: 31–44), Beyleveld (1991: 18).

[551] Cf. Gewirth's discussion of the role of the signalman, *RM*: 222–223.

[552] DSM-IV lists three: 'major depressive disorder', 'dysthymic disorder' and 'depressive disorder not otherwise specified' (1994: 318). ICD-10 includes 'depressive episode' and 'recurrent depressive disorder', and within that makes several distinctions relating to the severity of the condition, specifically whether or not it has somatic syndromes and whether or not it has psychotic features (1992: 28–29).

> "Five (or more) of the following symptoms have been present during the same 2-week period and represent a change from previous functioning; at least one of the symptoms is either (1) depressed mood or (2) loss of interest or pleasure. ...

1. depressed mood most of the day, nearly every day, as indicated by either subjective report (e.g., feels sad or empty) or observation made by others (e.g. appears tearful). ...
2. markedly diminished interest or pleasure in all, or almost all, activities most of the day, nearly every day (as indicated by either subjective account or observation made by others)
3. significant weight loss when not dieting or weight gain ... or decrease or increase in appetite nearly every day.
4. insomnia or hypersomnia nearly every day
5. psychomotor agitation or retardation nearly every day (observable by others, not merely subjective feelings of restlessness or being slowed down)
6. fatigue or loss of energy nearly every day
7. feelings of worthlessness or excessive or inappropriate guilt (which may be delusional) nearly every day (not merely self-reproach or guilt about being sick)
8. diminished ability to think or concentrate, or indecisiveness, nearly every day (either by subjective account or as observed by others)
9. recurrent thoughts of death (not just fear of dying), recurrent suicidal ideation without a specific plan, or a suicide attempt or a specific plan for committing suicide."[553]

The psychiatrists Goodwin and Ghaemi elaborate on this:

> Mood is bleak, pessimistic and despairing. A deep sense of futility is often accompanied, if not preceded, by the belief that the ability to experience pleasures is permanently gone. There is a slowing or decrease in almost all aspects of emotion and behaviour: rate of thought and speech, energy, sexuality and the ability to experience pleasure.[554]

Under precautionary reasoning, we should treat adults experiencing depression as ostensible agents in the vast majority of cases. In the most severe cases, adults experiencing depression will appear ostensibly as intermittent agents if the fluctuation and severity of mood inherent in the disorder gives rise to exhibiting rpvp at time t_1 and not exhibiting it at time t_2. To be clear about this, it is helpful to draw some distinctions. Mild or moderate experiences of the above symptoms will not influence having or displaying rpvp or the range of decisional or task competences that the person exhibits. There is also unlikely to be sufficient impairment of social functions that would warrant treating the person as a societally incompetent agent. When the depressed person visibly retains the generic capacities of action behaviour (GCAb) that exceeds that required for agency and when her behaviour gives no indication of an erosion of ability at a specific task or a specific decision, we should not doubt the presence of that task or decisional competence. In other words, the existence of mild or moderate depression *per se* should not affect judgments about the task and decisional competences of an individual where these are apparently undiminished.

Extreme phases of severe forms of depression, such as manic or clinical depression with psychotic features, may debilitate motivation and induce feelings of

[553] American Psychiatric Association (1994: 327).

[554] Goodwin and Ghaemi (2000: 677).

worthlessness to such an extent that the ability of the sufferer to value proactively her purposes (i.e. to have rpvp) or to act so as not to cause unintended harm to herself or others is temporarily impaired.[555] During this time, the agent will lack societal competence. In such cases of severe depression, the erosion of task and decisional competences can also be significant. Sufferers may no longer be able to perform tasks that were previously an entrenched part of their personal, domestic or occupational life because of apathy or lethargy induced by the condition.[556] Despair, hopelessness and negative expectancy may occlude perception, which give rise to a 'depressive realism' that can diminish the capacity to recognise the value of trust in themselves and others.[557] An important issue also arises as to what extent the person's value structure has fragmented and the integrity of their pre-existing self has eroded.[558] In spite of the ramifications of this latter point for discussions of selfhood and personal identity, it need not bear on considerations of agency under Gewirthian theory. As we have already acknowledged, the relevant question in making ascriptions of agency is whether the entity can still act *simpliciter* rather than act in accordance with previously expressed values or within the terms of reference of a pre-existing self.

Whilst we must treat all individuals who are experiencing severe depression as agents under precautionary reasoning, our judgments about the task and decisional competences of those individuals will vary according to the impact of the condition on the performance of those competences. This will also lead one to draw inferences from the degree to which she displays rpvp in a standing fashion that bear upon the permissibility of the individual developing new task and decisional competences. Where an individual suffering from severe depression displays rpvp in a way that is less close to a standing fashion, we should treat her as a societally incompetent agent, with a restricted range of task and decisional competences. Since this approach is motivated by a concern to protect the generic rights of the

[555] Of course, in practice, distinctions between mild, moderate and severe depression will hinge upon the frequency, nature and degree of the symptoms that are apparent. A psychiatrist will therefore look first to the patient's condition in order to reach a classificatory judgment rather than looking first to an 'ideal-type' definition and considering how well the patient's condition fits with one of these. See World Health Organization (1992: 121).

[556] Bech (2000: 683–684). For a powerful part-autobiographic account, see Wolpert (2001).

[557] As discussed in Garrett (1994), who argues that the pessimism associated with depression is irrational, even if it leads to more accurate self-observation, because it undermines the ability to pursue the good life. This meshes with the discussion in Beyleveld and Brownsword that a 'dignified' character and 'dignified' conduct are conducive to acting in accordance with the PGC (2001: 138–141). This does not imply that people experiencing depression of whatever degree lack dignity, however.

[558] Which could gives rise to a 'multiplicity' of selves (several apparently complete selves residing within the same body), although this more likely within dissociative identity disorder (Radden, 1996: 38–57). For discussions of the broader issues surrounding multiple selfhood and personal identity, see Williams (1970), Parfit (1984, esp. Chapter 12), Glover (1988, Chapter 1), Wilkes (1988, Chapters 4 and 5) and Haksar (1991, Chapter 6).

depressed individual *qua* cognitively vulnerable agent, it would be a warranted act of paternalism.[559]

Much like judgments of agency competence can be made with regard to comparing behaviour over time following the onset of the disorder, one may be able to make a comparative diachronic judgment to contrast the level of a particular type of task or decisional competence at t_1 (before the condition) and t_2 (after the onset of the condition). This helps to identify where there has been a marked deterioration in a particular competence of this type. What is at issue is not the *quantitative* loss of task competence following the onset of the disorder, but the *qualitative* effect of this loss.

The degree to which a depressed individual appears incapable of performing or participating in the range of activities at which she had previously task or decisional competence will determine the extent to which she appears either a specific task or decisionally incompetent agent or a societally incompetent agent. Apparent incapability may be determined by empirical observation or by subjecting the individual to an assessment.[560] In any event, severely depressed persons whose task competence has been eroded to the point where they can no longer cope independently with routine interactional situations will need to be subject to some assessment of their abilities in order to provide an appropriate level of support in such situations.[561]

Precautionary reasoning also allows us to make judgments about the degree of societal incompetence of individuals with depression. Whether we should treat a depressed individual as having a heavily or marginally restricted ability to engage independently in routine interactional situations depends upon to what extent the depressed individual's GCAb above that required for agency has diminished from time t_1 to time t_2. Marginal diminution suggests that she is more likely to retain her ability to engage in routine interactional situations independently, unless of course, she already experienced societal incompetence before suffering from depression. Greater diminutions will usually suggest a heavily restricted ability to engage in routine interactional situations independently. To allow the individual to continue exercising certain specialised task and decisional competences on the assumption that they are intact risks the individual causing unintended harm to herself[562] or others.[563] This is particularly likely if the course of the diminution is rapid or if her GCAb fluctuates.

---

[559] Which calls to mind Wikler's claim that "the strongest argument for paternalism applies where the aim is to protect from harm rather than to ensure a benefit" (1979: 385).

[560] I discuss assessments of decisional competence in Chapter 6.

[561] In any PGC-compliant polity or one that sincerely aspires to protect human rights.

[562] For instance, lack of understanding or appreciation of the relevant circumstances of a decision could erode competence to make financial decisions and decisions about consenting to or refusing medical treatment, due to an extreme inability to concentrate, lack of energy or irrational feelings of self-reproach.

[563] A severely depressed psychotherapist, for example, may be overcome with nihilism and fatigue whilst conversing with a client if returning to practice before having recovered from the condition, thus causing a transference of those feelings onto the client at an especially vulnerable time. Similarly, a severely depressed bus driver may not care whether he stops at a red light, thus threatening the safety of his passengers, other road users and himself.

The approach suggested by applying precautionary reasoning does not correlate the agency status of the depressed individual to the intensity of depression suffered and nor does it assume that forms of depression which exhibit fluctuations in mood and affect necessarily erode rpvp. By the same token, precautionary reasoning would not make the loss of societal, task or decisional competences "an essential criteria of severity"[564] of depression, but it would allow inferences to be drawn from their loss that are relevant to how we should treat the agent. It is also important to remember that a physician may responsibly prescribe antidepressant drugs in order to control the most destructive aspects of the person's depressive behaviour,[565] on the proviso that this was motivated by concern for the patient's welfare rather than as part of a wider policy of insidious social control. Crucially, precautionary reasoning requires us to look beyond the effects of the condition to search for any evidence of action at all, whilst at the same time creating an opportunity to gain an enhanced insight into the debilitating nature of a particular case of depression through examining changes in task and decisional competence.[566]

Adults with Schizophrenia

Schizophrenia typically causes an individual to experience a transformation in their cognitive functions and experience of social reality.[567] Extreme forms of depression and bipolar disorder share some of these symptoms, but in schizophrenia, these are typically more pronounced. DSM-IV gives the following definition of schizophrenia:

> "Two (or more) of the following, each present for a significant portion of time during a 1-month period (or less if successfully treated):
>
> 1. delusions
> 2. hallucinations
> 3. disorganized speech (e.g., frequent derailment or incoherence)
> 4. grossly disorganized or catatonic behaviour
> 5. negative symptoms, i.e., affective flattening, alogia, or avolition"[568]

Most experiences of schizophrenia are such that the force of misperceptions of reality, which includes false beliefs about oneself, delusions about the world and significantly fluctuating desires,[569] heavily compromise the capacity for rational understanding.

---

[564] The ICD-10 guidance cautions against incorporating 'social performance' in these criteria. See World Health Organization (1992: 121).

[565] This could offset the impact upon societal, task or decisional competences.

[566] Rudnick (2002: 153) endorses this comparative approach in relation to the preferences of the depressed person.

[567] Birchwood and Jackson (2001: 3).

[568] American Psychiatric Association (1994: 285). The DSM-IV criteria are more stringent than the ICD-10 classification, which requires the persistence of symptoms continually over six months (Birchwood and Jackson, 2001: 3). The 'sub-types' of schizophrenia include paranoid, disorganized, catatonic, undifferentiated and residual types (American Psychiatric Association (1994: 286–290).

[569] American Psychiatric Association (1994: 274–275).

Motility may be impaired through catatonic motor dysfunction (such as posturing, rigidity or excessive movement),[570] which in turn undermines the continuity of task competences. Withdrawal, self-neglect and extremes of behaviour can undermine social functioning and the performance of both task and decisional competences. As in cases of manic depression and bipolar disorder, this may be sufficiently great so as to evince ‘separate selves’, where the sufferer exhibits character traits and dispositions that are so inconsistent with the sufferer’s pre-existing personal identity that a new self or selves apparently emerge to rival the former, or to succeed it entirely.[571]

One of the greatest challenges to the pursuit of freely chosen purposes whilst suffering schizophrenia is the experience of delusions and auditory and visual hallucinations.[572] These often give rise to deluded premises for action (e.g. others intend to persecute or harm that person), false beliefs about the nature of mind and intentionality (e.g. that others can read the person’s mind, or that thoughts have been inserted into her mind), false perceptions of cause, motive or effect (e.g. that television or the Internet is sending messages informing the person how to behave or is controlling their behaviour), or beliefs of grandeur or invulnerability (e.g. that one is Jesus Christ, that one is being spoken to by God or that one will recover from a life-threatening illness through divine intervention). In extreme cases, the consequences of acting on these deluded premises could involve physical violence, although in most cases the sufferer directs this towards herself, rather than others.[573]

In light of these extraordinarily devastating symptoms, one of the main challenges for psychiatry is how we might go about *understanding* schizophrenia. Karl Jaspers, a pioneer of philosophical psychiatry, conceded that schizophrenia was inscrutable, deeming the condition “incomprehensible” and “closed to empathy”.[574] Such views allowed medicalized interpretations of the condition to thrive. Yet fifty years later, Anglo-American commentators were proposing a radical challenge to conventional psychiatric thinking about schizophrenia. For very different reasons, so-called ‘anti-psychiatrists’ Ronald Laing[575] and Thomas Szasz[576] both considered schizophrenia to be a form of self-expression in which conventional psychiatric practice should not intervene. Laing saw schizophrenia as representing an attempt to live through an unbearable situation. Rather than a manifestation of mental disorder, schizophrenia presents an opportunity for personal growth that liberates the sufferer from oppressive family relationships, in which we can find the origins of the condition.[577]

---

570 American Psychiatric Association (1994: 276).

571 See the discussion by Radden (1994, Chapter 4, esp. pp. 61–66).

572 Auditory and visual hallucinations are the most common, although olfactory (smell-related), gustatory (taste-related) and tactile (touch-related) varieties may occur.

573 Allebeck et al. (1986), discussed in Birchwood and Jackson (2001: 6).

574 Quotes attributed to Jaspers are taken from Read (2001: 449). The work for which Jaspers is renowned in this area is his *General Psychopathology* (1913, reprinted in two volumes, 1997a, b).

575 Laing first expresses his ideas on schizophrenia in *The Divided Self* (1960, repr. 1990).

576 Szasz’s seminal work is *The Myth of Mental Illness* (1974, repr. 2003).

577 This was the uncompromising position taken in the earlier of Laing’s writings, before *The Politics of the Family* (1972).

Szasz, alternatively, views the notion of mental illness as pathologizing voluntary requests for help in living, whilst providing an artifice for offenders to evade criminal responsibility.[578] On this account, once people suffering from schizophrenia are labelled mentally ill, they become the helpless victims of psychiatric power and a coercive state.[579] More recently, philosophers have turned their focus to the language of schizophrenia, disagreeing on whether it is a form of "hyper-rationality"[580] or simply "nonsense".[581] The world of the schizophrenia sufferer as interpreted from a Gewirthian standpoint may be similarly unintelligible, as Gewirth's theory of mind relates to what it means to be an agent rather than how we can make sense of mental illness. Nevertheless, we can still look for evidence of the ability to form a proactive evaluative relationship between oneself and one's own purposes (rpvp) amongst the solipsistic "mass of contradictions"[582] of schizophrenia.

However florid the symptoms are, it is highly likely that the schizophrenia sufferer will retain the capacity to value purposes, even if she perceives someone or something else to have given them to her, or they are in a constant state of flux. The difficulty arises when we consider in what sense we may take them to be the agent's own voluntary purposes. Purposes that the sufferer apparently values under the influence of a schizophrenic delusion (e.g. walking around the streets at night in order to decode messages she believes are being communicated to her through the registration numbers of parked cars) will be different from those valued during a period of lucidity. This alone may be insufficient to impair the exercise of certain task and decisional competences (in the example above, we assume that the sufferer is still able to find her way around a residential area, or distinguish a car registration number from the manufacturer's badge) but the motive behind exercising them is deeply irrational.

Through the application of precautionary reasoning, we can find a resolution to the problem of deluded purposes, at least on a moral level.[583] As we already have seen, denial of agency is the most serious violation of the PGC and we must treat any evidence that it may be present as persuasive, even where the entity in question is constrained in its exercise of agency.[584] In one important sense, there is no difference between a case where physical coercion is currently undermining the voluntariness of an individual's actions and a case where mental disorder has the same effect. In both cases, the force currently undermining the individual's

578 Reznek (1991: 94). Szasz presents this argument most forcefully in *Law, Liberty and Psychiatry* (1974).

579 Szasz, *ibid.*

580 Sass (1995). Sass's approach draws heavily on the ideas of Wittgenstein (1953, repr. 2002).

581 Read (2001: 456).

582 Read (2001: 469).

583 Precautionary reasoning cannot address the *ontological* significance of delusions and hallucinations for the schizophrenia sufferer. For an approach that seeks to consider this, see Phillips (2003).

584 Of course, this evidence may need to be interpreted in accordance with the criterion of avoidance of more probable harm.

voluntariness does not necessarily extend to negate their having the dispositional features of rpvp.[585]

To illustrate this, consider the following analogy. A person who a gunman threatens on the pain of death to do his bidding could not be understood as having lost her agency status whilst complying with his demands (even though we would typically absolve the victim of responsibility for any ensuing harm). It would be more appropriate to think of the coercion as having overwhelmed the agent's current capacity for free choice.[586] Similarly, schizophrenia manipulates desires and preferences but does not necessarily eliminate the dispositional features of rpvp, or the metacognitive capacity to know that the condition is manipulating one's desires and preferences.[587] It would be more helpful to think of the schizophrenia as having driven the exercise of rpvp into abeyance whilst the disposition for rpvp endures. Therefore, for as long as there remains a possibility that the sufferer retains the dispositional features of rpvp, we should treat the sufferer as an agent, albeit one with restricted task and decisional competences.

Since the symptoms of schizophrenia tend to vary in proportion to the severity of the condition, we can identity a sliding scale along which there is a *prima facie* correlation between its severity, the display of rpvp and the display of task and decisional competences. Where the condition fluctuates, it is possible to suggest that sufferers of schizophrenia are, like persons experiencing depression, either ostensible agents or ostensible intermittent agents depending upon whether the display of rpvp at time t_1 and at time t_2 represents a frequent occurrence for all instances of such observation. This, combined with understanding of the pathogenesis of schizophrenia, yields a precautionary response along the following lines. In mild cases of schizophrenia, we should assume that rpvp is retained along with most task and decisional competences. In moderate cases of schizophrenia, we should assume that rpvp is retained, but that it is highly probable the effects of the condition will undermine task and decisional competences previously held and may warrant treating the sufferer as a societally incompetent agent.[588] In severe cases of schizophrenia, we should still assume that the sufferer retains rpvp, but there may be extensive periods when it is not apparent and when most task and decisional competences (and the ability to develop new ones) are apparently lost.[589] Such sufferers will manifestly lack societal competence.

There are two caveats that apply to this. First, whether or not a schizophrenia sufferer is taking medication for her condition will not affect the precautionary

---

[585] See the discussion in *RM*: 28–35.

[586] Compare Gewirth's discussion of the gunman scenario, *RM*: 32.

[587] See Breier and Strauss (1983).

[588] This is because when in the throes of schizophrenic delusion, the sufferer's ability to interact independently in her environment is impeded to an extent that warrants paternalistic intervention (as measured by the PGC).

[589] In these severe cases, there may be a profound dissonance between the display of rpvp and the having of rpvp (although of course we can never prove this beyond doubt).

grounds for making a judgment about her agency. We should presume that she has a disposition for rpvp which endures irrespective of her taking medication.[590] However, the knowledge that a schizophrenic sufferer (or, for that matter, any other mentally disordered person) uses medication to sustain mental equilibrium might affect whether we treat that person as a societally competent agent and as having competence at particular specialised tasks and decisions. Should the medication improve the sufferer's independent functioning in society, the less likely it will be that we should treat her as a societally incompetent agent. It may also allow for the return of task and decisional competences previously lost. Alternatively, the medication might impair the sufferer's array of prospective or developable task and decisional competences if its effects make the development or exercise of such a competence difficult or potentially harmful.[591] It will also inform our expectations of how we may have to treat the individual at a future date if her condition persists and she is no longer taking medication.

Second, the retention of task and decisional competences will be more evident in situations when sufferers experience an isolated, mild episode of schizophrenia, or are in advanced states of recovery. In the case of an isolated episode, task and decisional competences are likely to return, even if they were lost during the height of the condition.[592] The individual's schizophrenic episode should have no continuing bearing upon presumptive or demonstrable task or decisional competences as it no longer jeopardizes them and is unlikely to do so in the future. In the case of recovering schizophrenia sufferers, the consistency with which the sufferer displays the generic capacities of action behaviour consistent with rpvp (GCAbrpvp) typically will be greater than an individual who is not in remission, although it will take some time for the individual concerned to have rpvp restored as an entrenched feature of her personality. In many cases, therapy or medication will help to prevent relapse[593] and will offer a means to reclaim the constant ability to exercise rpvp. Once the sufferer has made consistent progress towards healthy functioning (at least sufficient for clinical recovery[594]), she will be ready for reintroduction to

[590] Although this should not imply that agency under the PGC entails a 'natural' state. A Gewirthian definition of agency can accommodate the possibility of rpvp brought about through artificial means. For instance, a cyborg created by human beings would be an agent, even though this agency has not arisen organically. On this account, it is irrelevant to the task of identifying evidence for agency how rpvp has emerged or how it is sustained. The origins of agency are only morally relevant if it has come about at the expense of violations of the generic rights of other agents. Agent-entities that are the product of a prior infringement of another agent's generic rights, such as a cyborg derived from the brain tissue of an unwilling donor, would be an example of this.

[591] For example, medications that interfere with the ability to operate machinery or drive a motor vehicle would impair associated task competences.

[592] Jablensky (2000: 613–614).

[593] See respectively, Chadwick et al. (1996), Leff and Wing (1971, discussed in Birchwood and Jackson, 2001: 101–102).

[594] Warner (2003) identifies clinical recovery as a precondition of social and psychological recovery.

everyday activities and situations in order to restore her lost independence. This would require a therapeutic environment to nurture autonomy through revivifying pre-existing or latent talents and abilities and the exercise of rational judgment.[595]

Adults with Alzheimer's Disease and Other Dementias

Dementia is a progressive, usually irreversible condition that causes cognitive vulnerability through degeneration in higher cognitive functions, especially memory, reasoning, comprehension and judgment. It particularly affects individuals above sixty-five years of age, although some sufferers are diagnosed with 'early-onset' dementia.[596] Alzheimer's disease is the most common form of dementia,[597] but dementia can also arise in Pick's disease, Creutzfeldt-Jakob disease, Huntingdon's disease, Parkinson's disease or through vascular dementia.[598] In all dementias, especially Alzheimer's disease and Pick's disease, changes in the structure of the brain become apparent, such as neuronal depletion, a reduction in neurotransmitters and enzymes, along with lesion development, particularly extracellular amyloidal plaques and intracellular neurofibrillary tangles.[599]

DSM-IV defines dementia, which includes Alzheimer's disease, in the following terms:

> The essential features of a dementia are multiple cognitive deficits that include memory impairment and at least one of the following cognitive disturbances: aphasia [impairment of language functions], apraxia [impaired motor abilities], agnosia [failure to recognise objects], or a disturbance in executive functioning [such as abstract reasoning]. The cognitive deficits must be sufficiently severe to cause impairment in occupational or social functioning and must represent a decline from a previously higher level of functioning.[600]

These 'multiple cognitive deficits' have enormous implications for the sufferer. Soon after onset, independent life becomes progressively difficult to maintain due to neuronal loss in the hippocampus (the part of the brain responsible for transforming

[595] Ideally, such as a 'therapeutic community'. These 'communities' would be run by a collective of mental health professionals and current and former sufferers and funded by central government. They would be antithetical to the practice of consigning recovering schizophrenic sufferers to flawed 'care in the community' programmes of the type that were introduced in Britain from the 1960s.

[596] American Psychiatric Association (1994: 140). Cases involving people below fifty years of age are much rarer (1994: 141).

[597] Lovestone (2000: 390). Alzheimer's disease accounts for between 60% to 70% of all incidences of dementia. Over 20% of individuals above eighty-five years old suffer from Alzheimer's disease (*ibid.*: 391).

[598] The ICD-10 classification also includes dementias of an unspecified type, such as senile dementia.

[599] World Health Organization (1992: 47).

[600] American Psychiatric Association (1994: 134).

short-term memory into long-term memory).[601] Short-term memory and language skills are usually the first to decline, leading to visuospatial and motor impairment. At the same time, skills of information assimilation and recall deteriorate, beginning with new information, but spreading to affect entrenched knowledge and understanding in the more advanced stages. This impairs concentration, logical reasoning, decision-making and communication. Throughout, consciousness remains intact, which means that the sufferer is painfully aware of her slow deterioration.[602] Most tragic of all are changes in personality. The ability to engage previously held values, character traits and a conception of the good diminishes gradually and steadily. This leads to the unravelling of selfhood in a process of effacement or "unbecoming",[603] captured poignantly in the reflections of a sufferer: "Every few months I sense that another part of me is missing. My life ..., my self ... are falling apart. I can only think half thoughts now."[604]

Inevitably, this systematic disintegration of higher-level cognitive functions impinges upon autonomy and agency. In its early stages, the effect confines itself primarily to the loss of task and decisional competences, spreading progressively through the middle stages to eventually undermine the success with which one can act for freely chosen purposes at all. The extent to which agency survives the mental ravages of severe dementia is a topic that has generated much interest in bioethics and philosophy of mind. A central question is whether the effect of personality change on a sufferer's identity is qualitative only, or sufficiently great as to give rise to a numerical change in the sufferer's identity so that she is, essentially, a different person from the one she was.[605]

Two approaches structure the terms of this discussion. The first is the view initially posited by John Locke and revised by Derek Parfit that psychological continuity and connectedness are constitutive of personal identity.[606] Psychological connectedness involves the possession of directly connected relationships with past, present and future selves whereas psychological continuity is the possession of "overlapping chains of strong connectedness".[607] For Parfit, connectedness is more important than continuity. He supports a variety of reductionism, in which the 'R-relation' of connectedness displaces reliance on some 'further fact' about

---

[601] Geula (1998), discussed in Jaworska (2006: 94).

[602] Illustrated with great pathos in the film *Iris* (BBC Films, 2001), a dramatisation of the philosopher and novelist Iris Murdoch's life and eventual affliction with dementia.

[603] Fontana and Smith (1989). See also the discussion in Post (1995).

[604] Excerpted in Radden (1996: 276; originally quoted in Cohen and Eisdorfer, 1986: 21).

[605] McMillan (2006: 64).

[606] Presented in Locke (1997, originally published 1689) and Parfit (1971, 1984). There are notable differences between the approaches of these two philosophers, particularly Parfit's rejection of Locke's use of memory as a criterion of personal identity, which he believes begs the question in favour of personal identity (1971: 16). Parfit's response is to substitute the memory criterion with a notion of '*q*-memory' (1971: 14–15).

[607] Parfit (1984: 206).

psychological existence, such as having a soul or Cartesian ego.[608] On this account, the dementia sufferer, especially in the more advanced stages, would represent a numerically different person inhabiting the same body as her former self once did.

The second approach to have informed the discussion I will call the mixed account. Proponents of such an approach implicitly or explicitly associate identity with brain function *and* spatiotemporal embodiment. These either tend towards 'a common sense' view of personal identity in which the physical unity of our brains and bodies determine our identity[609] or towards 'animalism',[610] which argues that our identities are determined by species membership alone (e.g. being a human animal) rather than by any psychological property of being human. On this basis, we can explain radical change, including dementia, as representing a different stage within the life of the same person, rather than as a numerically different person inhabiting the same body. Of the accounts of the relationship between identity and dementia to have emerged in recent years, several display an affinity with the mixed account.[611]

Ronald Dworkin, in *Life's Dominion* and in an earlier paper, 'Autonomy and the Demented Self', presents such an account that focuses upon the impact of dementia on autonomy. In Dworkin's view, "personal identity does survive even the most serious dementia".[612] Dworkin's theory rests upon an assumption that the psychological discontinuity between a human being with agency competence and the same human being after the loss of agency competence induced by dementia does not amount to a rupture of personhood. For the same reason, the psychological discontinuity of dementia does not undermine the moral significance of the sufferer's previously expressed values and wishes.[613]

Dworkin distinguishes between two types of interests he believes human beings have – critical interests and experiential interests – in order to elicit what is significant about loss of autonomy in dementia. Critical interests emanate from a person's value framework or conception of the good that has endured over time. They are distinctive to cognitively functional human beings and generate a sense of narrative coherence within one's life. By contrast, all sentient beings possess experiential interests, which derive from a transient desire for pleasure satisfaction and pain avoidance.[614] Of the two types of interest, critical interests epitomise autonomous

---

608 Parfit (1984: 206). See also Korsgaard (1989, esp. pp. 103–109) for a discussion of reductionist and non-reductionist accounts of personal identity.

609 See Williams (1973) for a leading account.

610 A recent approach developed by Olson (1997).

611 For a conspectus of recent approaches, see Hughes et al. (2006).

612 Dworkin (1986: 6). Dworkin has also reaffirmed this view more recently in a reply to critics, stressing that he takes competence and dementia to be different stages in the history of the same person, who remains "the same, single object of ethical inquiry" (2004: 367).

613 Dworkin (1986: 5). Dworkin makes it clear that in discussing dementia he is referring principally to the loss of agency competence rather than task or decisional competence (1994: 225).

614 Dworkin (1994: 201–204).

action. Dworkin argues that a dementia sufferer will have possessed both critical interests and experiential interests before the onset of dementia but that she can only sustain experiential interests in the most acute phase of her cognitively vulnerable state.[615]

Dworkin conceives of autonomy as serving to uphold the integrity rather than the welfare of the agent in question.[616] This entails the ability "to act out of genuine preference or character or conviction or a sense of self".[617] However, a purely integrity based account of autonomy would deny dementia sufferers a right to autonomy since, in Dworkin's view, they lack the occurrent capacity for autonomous action.[618] Some modification to this account is therefore required. In order to respect the autonomy right of the dementia sufferer *qua* previously autonomous person, Dworkin argues that we must respect her earlier critical interests in favour of her current experiential interests. He refers to this as the 'precedent' doctrine of autonomy.[619] Dworkin does emphasise, however, that this does not mean that there will never be good reasons for acting in accordance with her experiential interests, but if we do act in this way, we must accept that we are necessarily infringing her autonomy.[620]

Dworkin's account of critical and experiential interests rests upon a controversial demarcation between an agentive and a post-agentive state in dementia. That he does not argue for this demarcation and instead assumes it makes his account all the more susceptible to criticism. Agnieszka Jaworska argues that Dworkin pitches the abilities required for agency competence too high.[621] For Jaworska, the essence of autonomy is the capacity to value construed in a minimal sense.[622] Valuing, for Jaworska, comprises thinking one is right to want what one wants, an association between the attainment of the desired goal and one's sense of self-respect and independence between the significance of pursuing of the goal and one's own experience.[623] On this basis, she distinguishes the ability to value from the ability to understand the narrative of one's life.[624] Dworkin's understanding of autonomy fails to capture this

---

615 Dworkin (1994: 225).

616 Dworkin (1994: 223–224).

617 Dworkin (1994: 225).

618 Dworkin (1994: 225–226).

619 Dworkin (1994: 226–229).

620 Dworkin (1994: 229). Dworkin's suggestion that it might be "morally unforgivable not to try to save the life of someone who plainly enjoys her life" (1994: 228) seems to me a compelling reason not to honour the person's previously expressed wish to refuse treatment, at least in circumstances where pain and suffering does not outweigh this enjoyment. Dresser (1995) appears to share this view.

621 Jaworska (1999, 2006).

622 Jaworska (1999: 116).

623 Jaworska (1999: 116). An example of the independence between the importance of the goal and one's own experience would be where one values the end for reasons unrelated to any improvement in one's state of mind, such as undertaking low paid menial work (which would give rise to no pleasure or improvement in one's state of mind) in order repay one's student debts.

624 Jaworska (1999: 124).

fundamental first sense of valuing, instead focusing upon the second more demanding sense of valuing in the context of one's own life's projects and commitments.[625]

Jaworska argues that dementia sufferers, especially those yet to reach the advanced stages of the condition, are capable of this less demanding form of valuing. She justifies this claim with reference to recent research in neurophysiology. This suggests that dementia leaves intact those areas of the brain which are primarily responsible for generating value until the most advanced stages.[626] The primary effect of damage to the hippocampus – which is where the damage of the earlier stages is concentrated – is to render the person "unable to update her autobiographical narrative",[627] rather than unable to value. By way of illustration, Jaworska offers vignettes featuring dementia sufferers who demonstrate the metacognitive capacity to be aware of the effect of their decline as it is happening to them.[628] According to the minimal valuing hypothesis, valuing occurs even when it takes the form of lamenting the loss of abilities one no longer can exercise.[629]

Jaworska believes that if Dworkin's account of critical interests is reframed as nothing more than "opinions about what is good for me",[630] then it can be read as supporting a notion of valuing which connotes a sense of the agent's ability to make judgments of value in the present, rather than judgments of value that are informed by a narrative self-understanding. What is important according to this revised account of critical interests is that we respect the 'contemporaneous autonomy' of the dementia sufferer *qua* occurrent valuer rather than her precedent autonomy.[631] This involves enhancing the sufferer's autonomy through support, assistance and dialogue[632] and may require preparedness to disregard the sufferer's apparent choice where this decision conflicts with her more fundamental occurrent values.[633]

Others have offered parallels to Jaworska's argument. In a critical review of *Life's Dominion*, Rebecca Dresser questions the extent to which one may draw a clear boundary between critical and experiential interests.[634] For Dresser, one can rationally choose a purpose precisely for its immediate or sensuous benefit.[635] She takes a

---

[625] Jaworska (1999: 130). Shiffrin (2004) takes a similar view in relation to the exercise of autonomy: "[w]e show respect for the demented by allowing their contemporary interests and their voiced will to exert substantial influence on our deliberations" (*ibid.*: 213).

[626] Jaworska (1999: 122). Her principal sources are Damasio (1994) and Braak and Braak (1995).

[627] Jaworska (1999: 121).

[628] Jaworska (1999: 105–108).

[629] Jaworska (1999: 114).

[630] Jaworska (1999: 113; quoting Dworkin, 1994: 202).

[631] Jaworska (1999: 133–134, 2006: 99).

[632] Jaworska (1999: 126, 133–134).

[633] Jaworska (1999: 135–136, 2006: 99). The scope for overriding the apparent choice of the sufferer for the sake of her autonomy interests bears similarities with Feinberg's notion of soft paternalism (1986: 12).

[634] Dresser (1995).

[635] Dresser (1995: 36).

sceptical view towards whether there is a general preference to live one's life according to a narrative structure, and instead suggests that many human beings evince a willingness to adapt to circumstances as they present themselves.[636] Unlike Jaworska, however, she concedes that a Parfitian account of personal identity (which Dworkin rejects) may be plausible.[637] Julian Hughes, alternatively, supports the Dworkinian view of personal identity,[638] and sketches a philosophical account of personhood in dementia that dovetails with what Hughes claims is the "reality of clinical experience".[639] He argues for a 'situated-embodied' view of the agent, in which the agent is constituted by her physiological, social and cultural context – her "embeddedness and … embodiment".[640] On this account, personhood (which Hughes uses interchangeably with agency) persists throughout dementia until the point of cardiac arrest.[641] To act on this premise requires that we should encourage the agentive abilities of the sufferer for as long as possible.[642] It follows that we are under a moral obligation to assist a dementia sufferer in the continuance of her agency.

Viewed together, Jaworska, Dresser and Hughes lend weight to the argument that the dementia sufferer – at least in the earlier stages of the condition – is still a valuer. In particular, the positions of Jaworska and Hughes appear to support an approach compatible with precautionary reasoning. Their respective ideas of construing valuing minimally and encouraging agentive capacities meshes with the Gewirthian position in two important ways. First, they emphasise that an agent is an entity that acts for any purposes it has freely chosen, whatever these might be, rather than specifically within the context of a self-generated narrative. Second, they reaffirm the positive right to assistance in having the basic goods of the PGC when one cannot achieve these through one's own efforts. This underpins the presumptive judgment under precautionary reasoning that a sufferer of Alzheimer's disease is ostensibly an agent until the GCAbrpvp ceases to be evident. Given the inevitability of such deterioration, there will come a point in the pathogenesis of every sufferer where she ostensibly ceases to be an agent and ostensibly becomes

---

636 Dresser (1995: 35).

637 Dresser (1995: 35).

638 Hughes (2001: 80).

639 Hughes (2001: 86). Hughes emphasizes that that his use of 'experience' is not empirical; rather it presupposes metaphysical conceptualisation in a Kantian sense (*ibid.*: 87).

640 Hughes (2001: 87–89). Similarly, McMillan (2006) argues from a starting point of situated agency but differs from Hughes in that he suggests that the effects of Alzheimer's disease prevents the person so afflicted from engaging in the 'webs of interlocution' in which personal identity is formed (2006: 70). Even if the person is numerically the same, we might have reason to believe that she has experienced profound qualitative change. This understanding may have the effect of eroding the obligations that family and friends previously felt towards that person (2006: 69–70).

641 Hughes (2001: 89).

642 Hughes (2001: 90). Such assistance may compensate for the growing powerlessness encountered in dementia. As Kitwood (1997) puts it, there is "a vast difference between a tragedy in which persons are actively involved and morally committed, and a blind and hopeless submission to fate" (1997: 69).

an intermittent agent, and then ostensibly ceases to be an intermittent agent and apparently becomes a partial agent.[643] However, we should not forget that this is an *evidential* property of sufferers; precautionary reasoning would reject a knowable Dworkinian division of agency and post-agency.

As with intellectual disability, it is helpful if we divide the severity of the condition into different levels. Under precaution, we should treat individuals who are mildly afflicted with the disease or who are in the early stages of its development as retaining agency competence but with decreasing societal competence. At this stage, it would be wise to keep existing task and decisional competences under review should one or more of them suddenly fail.[644] There should still be a presumption that sufferers in the middle stages of the condition be treated as ostensible agents, albeit with considerably impaired task and decisional competences, due to the damage to retention and recall of information.

It is important at this stage for health care professionals and carers to engage in activities that would assist in developing what remains of the sufferer's capacity for independent decision-making,[645] but it may be necessary to treat the sufferer (for her sake) as a societally incompetent agent. By contrast, individuals in the most advanced stage of dementia will appear to have lost the occurrent ability to exercise rpvp, at least most of the time. We can position such advanced dementia sufferers on a continuum somewhere between ostensible intermittent agency and apparent partial agency. In these cases, attempts to enhance the sufferer's autonomy would be likely to fail or seem clearly futile. Our duties of protection to them therefore should be paramount.

Summary

Under a precautionary application of the PGC, there is no doubt that we must treat the individuals in all five of these cognitively vulnerable groups as agents of some kind. From the available psychiatric evidence, the presence of each of these forms of cognitive vulnerability will rarely, if ever, erode agency competence. Only in the severest cases where those cognitive functions which sustain the ability to form a proactive evaluative relationship between oneself and one's own purposes (rpvp) are likely to have atrophied (such as in the most advanced stages of dementia) or

---

[643] This final stage of severity may not be reached in all cases of dementia, as there is a possibility that death may occur first.

[644] For example, such as where an individual who has been able to make independent decisions about her financial affairs suddenly starts to claim that she can no longer remember where she has invested her money, or starts buying a week's worth of groceries several times each week because she forgets that she has already bought the groceries for that week.

[645] Kitwood (1997) provides a number of examples of these.

have never developed at all (such as in the case of adults with profound intellectual disabilities) can we begin to consider that agency competence may be lacking. Even then, morally speaking at least, we cannot reach a judgment that they are not agents altogether – we still owe them extensive duties of protection under the PGC, based on the highly probable possibility of their being agents who cannot display rpvp.

A precautionary application of the PGC also must be the arbiter of any decision about when we should treat cognitively vulnerable individuals as agents of full or restricted societal competence and how we should treat cognitively vulnerable individuals whose means of display is limited. Under precaution, we should reach a decision about an agent's societal competence or incompetence based upon the consequences (as measured by the PGC) for the agent concerned and other agents of presuming one way or the other.[646] We must continue to assume the presence of agency competence even if the means of display is limited, such as in the case of motor neurone sufferers whose communicative abilities are significantly restricted in the absence of artificial means of display. In these cases, sufferers have a positive right to communicative assistance under the PGC.

The same precautionary approach can also resolve conflict between the institutional recognition of young people's societal competence on a case-by-case basis and the allocation of scarce resources to some other public good. In societies where administrative burdens are high and resources are scare, the directing of funds to support health and welfare initiatives is more important than directing them towards a case-by-case recognition of societal competence. This, however, should not undermine the arguments for a reconsideration of the threshold age at which institutional recognition of societal competence is set.

The presumption in favour of agency competence therefore does not extend in the same way to judgments of task or decisional competences. A denial of an ostensible agent's task or decisional competence is a serious matter, as it denies an agent's additive right to well-being, but it is nowhere near as serious as denying agency altogether, as this denies the entity in question the protection of the generic rights. A judgment of a continuing task or decisional competence or about the probability of its development will depend upon the level of the generic capacities of action behaviour (GCAb) that the agent exhibits above that required for rpvp (GCAbrpvp).

We have seen in a number of cases, especially in depression and schizophrenia, that the effect of an individual's cognitive vulnerability on her intellectual and conative function reduces her ability to exercise or develop task and decisional competences. Making a defensible judgment about the impact of the cognitive vulnerability on a particular task or decisional competence requires empirical evidence about the relationship between the cognitive vulnerability and performance of the task or decision concerned. If the empirical evidence should establish a

---

[646] This does not amount to understanding the PGC as a conventional consequentialist moral theory (which would seek to maximize consequences for goods that may be independent of agency rights), but instead as a form of "deontological consequentialism", in which the criterion of rightness for any action is the consequences for upholding the goods that are justified by the rights of agents. See *RM*: 216 and Beyleveld and Brownsword (2001: 184, n. 19).

likelihood of such competence being manifest among members of that cognitively vulnerable group, then we have an ethically defensible ground either for making a rebuttable presumption in favour of the presence of the competence or for channelling resources into nurturing its development. Conversely, if the evidence is doubtful, and we cannot undertake a case-by-case assessment of competence for the administrative reasons outlined above, then there should be a presumption against the presence of the competence and fewer resources directed towards its development. We can state this principle as follows:

Where X is ostensibly an agent and the available evidence suggests that X can exercise a particular task or decisional competence, X should be presumed able to do this. Where the evidence suggests that X can develop this competence with appropriate support, X should be presumed able to do this, and as far as possible, be given appropriate support. Institutional structures within a PGC-compliant polity (including the law) should permit the exercising or development of this competence unless the consequences of permitting X to exercise or develop this competence are worse (as measured under the PGC) than not permitting X to exercise or develop this competence.

This represents the normative foundation of a theory which I will call *precautionary task or decisional competence judgment* (PTDCJ). It is precautionary because it relies upon a precautionary application of the PGC. It refers to task or decisional competences in general terms, as we could apply the theory to a varied range of specific activities and decisions, such as learning to drive, voting in a general election, consenting to or refusing medical treatment or training to become a social worker. It also can be applied both to the formulation of a threshold test of competence (such as age) or to the formulation of a case-by-case determination (such as a test). That precautionary reasoning mediates between the application of ethical judgment and the assessment of the available evidence allows for a genuine 'reflective equilibrium' between the permissibility of the presumption and the possibility of the competence.

In the next chapter, we will apply PTDCJ using empirical studies in psychiatry to consider which, if any, members of these five cognitively vulnerable groups we should judge decisionally competent to consent to biomedical research and the grounds on which we can make a presumption in favour or against such competence.

Chapter 6
Cognitive Vulnerability and Consent to Biomedical Research

Whether we can make a judgment about whether an agent possesses the competence to make a decision, or even has developable ability to make that decision, requires empirical investigation as well as moral argument.[647] This requires us to consider relevant empirical studies into competence to consent to biomedical research. In this chapter, we will apply the theory of precautionary task or decisional competence judgment (PTDCJ) using the findings of psychiatric and psychological studies to establish which, if any, members of these five cognitively vulnerable groups we should judge competent to consent to biomedical research and the grounds on which we can make such a judgment.

It is helpful to begin by clarifying the meaning of 'judgments' of competence, as raised in the last chapter. We can do this by drawing a distinction between *assessing* and *judging* decisional competence. Assessing decisional competence is the process of evidence gathering about the decisional competence of a particular individual. A competence assessment tool, such as the MacArthur Competence Assessment Tool for Clinical Research (MacCAT-CR) or the Mini Mental State Examination (MMSE), will typically structure the form and substance of the information obtained. Judging decisional competence, alternatively, relates to what we do with this evidence. By this, we mean what normative significance we attach to it and how we use it to inform policy. Judging decisional competence requires a moral evaluation of the evidence generated by the competence assessment to form the basis for an ethical or legal norm.[648]

---

[647] A view also shared by Roberts (2000: 542).

[648] Kim et al. (2001: 713) suggest a different perspective on this distinction. For them, assessing decisional competence to consent can itself be thought of comprising an element of judgment, in which one measures relevant decisional abilities displayed by the decision-maker (against an appropriate standard) followed by a clinical judgment based on whether the decision-maker has demonstrated these abilities satisfactorily within the context of the decision to be made. The form of judgment described relates to an analytical type of professional judgment rather than a normative one premised on a moral theory. In subsuming the question of judgment within the matter of assessment, Kim et al. fail to give sufficient emphasis to this second type of normative judgment. This type of judgment is unavoidable for researchers and health care professionals when they are required to give reasons to explain a judgment of decisional competence or incompetence that go beyond recourse to a competence test score or some other quantitative indicator (e.g. when justifying a judgment of decisional incompetence to a patient's family members, etc.).

P. Bielby, *Competence and Vulnerability in Biomedical Research*,

Judgments of competence are rarely uncomplicated, however. The morally laden dimension subverts any claim to a value-free outcome. As Roth et al. observe, "[J]udgments of competency go beyond semantics or straightforward applications of legal rules; such judgments reflect social considerations and social biases as much as they reflect matters of law and medicine".[649] Whilst we cannot deny the social context in which judgments of competence take place, equally well, Roth's claim serves as a salutary reminder of the need to eliminate as far as possible unreflective or irrational value judgments about the circumstances in which one should deem an agent competent in respect of a particular decision or activity. As we have seen, the PGC offers a non-arbitrary starting point for making such judgments.

In this book, we are less concerned with how a researcher or physician might go about assessing competence or the content of a competence assessment test than with the ethical and empirical considerations that should inform a judgment that decisional competence is present or absent. This is not to separate the two matters entirely, nor to suggest that competence assessments are value-free, but to place our focus on the morally laden judgment dimension of competence determinations rather than the clinically laden assessment dimension.[650] Through an empirical investigation into the possibility of cognitively vulnerable individuals developing or exhibiting decisional competence to consent to research, we will be able to reach a more informed position under PTDCJ. Given the ultimate lack of certainty surrounding whether an entity we take to be an agent does actually possess a competence, this is the best we can aspire to. Before we turn to that, however, let us consider the central normative issue surrounding competence assessment that has a bearing on competence judgments.

Decisional Competence and Supported Decision-Making

Over twenty-five years ago, the President's Commission for the Study of Ethical Problems in Medicine and Biomedical and Behavioural Research, undertaken on behalf of the US government, claimed that an association between a person and a particular group of vulnerable persons should not alone determine the judgments we make about that person's decisional competence:

> The fact that a patient belongs to a category of people who are often unable to make general decisions for their own well-being … should alert health care professionals to the greater possibility of decisional incapacity. But it does not conclusively resolve the matter.[651]

---

[649] Roth et al. (1977: 280).

[650] Kopelman (1990: 309) claims that it is impossible for assessments of decisional competence to be value free. I also take this view. However, throughout her article, Kopelman appears to use 'assessment' and 'judgment' inter-changeably, which elides the important distinction between them.

[651] President's Commission for the Study of Ethical Problems in Medicine and Biomedical and Behavioural Research (1982: 171–172).

If we interpret the phrase "people who are often unable to make general decisions for their own well-being" to include at least some individuals with cognitive vulnerability, then it is likely that the Commission would maintain that cognitive vulnerability is a necessary but insufficient ground for making a judgment of decisional incompetence. In other words, the presence of a cognitive impairment in a potential participant may well incline a competence assessor to seek to investigate decisional competence, without it constituting enough reason on which to reach a judgment that decisional competence is absent. Successive commentators have echoed this point in recent years to the extent that it now seems rather trite.[652]

However, if we are to be guided by PTDCJ, the process of competence assessment must not only respect but also enhance the participant's autonomy *qua* agent. As we have seen, the PGC requires that we give other agents assistance in obtaining the generic goods when they cannot obtain them through their own unaided action. It is therefore vital that the competence assessment process amounts to more than a mere detached observation of a potential participant's decision-making abilities. It should represent an opportunity to empower the potential participant through presenting as much information as possible in a manner which she finds most accessible. The aim should be to elicit any latent decision-making skills of the individual through dialogue and encouragement. We cannot claim to have made a judgment of incompetence in good faith until this process has been undertaken. This is particularly salient when considering the decisional competence of cognitively vulnerable individuals, for whom the presence of support in the assessment process may make the difference between a judgment of decisional competence or incompetence. On this point, I agree with Freedman, who notes:

> [The] burden of informing the person is ... a heavy one. If understanding cannot be achieved by the ordinary process, then further methods must be sought. The method of informing chosen must be one designed to enlighten the individual in question. Only when it is clear that no manner of informing the patient will serve does the issue of competence become moot.[653]

With that in mind, it is helpful to consider what type of interventions might improve decisional competence prior to making judgments about the presence or absence of competence in a particular case. These need not be narrow in scope. Interviews, feedback, highlighting prior misunderstanding, and the use of repetition as a means of compensating for impaired attention represent common approaches to improving an individual's decisional competence to consent or to refuse biomedical research participation.[654] I will refer to such methods generally as 'supported decision-making'.[655]

One recent study has already shown these interventions and others to have had some success. Flory and Emmanuel surveyed a selection of research projects conducted

---

[652] See, in particular, Lo (1990) and Sullivan and Younger (1994). For an argument along the same lines which prefigures the President's Commission, see Baumgarten (1980).

[653] Freedman (1981: 65).

[654] Dunn and Jeste (2001: 598).

[655] I present an argument for supported decision-making with adult cognitively vulnerable research participants in Bielby (2008, forthcoming).

between 1966 and 1994, which compared enrolment through a standard informed consent process and enrolment using efforts to improve patient understanding.[656] These included multimedia, enhanced consent forms, counselling and 'neutral educators' across a range of participant populations, including people with mental disorders and psychiatric patients. Of the efforts undertaken in order to improve patient understanding, the authors found that time spent talking on a one-to-one basis (whether through neutral educators or from a member of the research team) was proven to be the most effective way of raising potential participants' understanding about the nature of the research in which they had been invited to participate. They concluded that "direct human contact tends to be more successful in improving understanding … [this] has more potential for active engagement and responsiveness to the individual needs of the research participant."[657]

Supported decision-making dovetails with a growing view on decisional competence assessment which emphasizes that what is important is not how competence is assessed but how decisional competence can be developed through "interaction and dialogue".[658] The more that we treat competence as a dynamic (developable) rather than a fixed (given) quality of an agent (where the evidence suggests that the competence can develop with appropriate support), the more scrupulous a competence assessment must be and the more ethical is our resulting judgment of competence or incompetence.[659] This is because a rigorous assessment process will yield more evidence on which to base a judgment of competence or incompetence, reducing the margin of error and the potential for 'social bias'. In supported decision-making, therefore, we can identify the normative intersection between assessing and judging decisional competence, which establishes a basic precondition for drawing any morally defensible competence judgments.

---

[656] Flory and Emmanuel (2004). Eyler and Jeste (2006) provide a similar survey of supported decision-making initiatives.

[657] Flory and Emmanuel (2004: 1599). In one study reported in this article, Stiles et al. (2001) found that, contrary to expectations, neither a graphically enhanced consent form nor a third-party facilitated consent process had a noticeable effect on improving comprehension among participants with depression, schizophrenia or among a control group. Indeed, the authors claim that the intervention of the facilitator may have instead confused some of the individuals with schizophrenia (*ibid.*: 784). However, the provision of iterative feedback, which involved the research assistant reading the consent form aloud followed by an opportunity for questions and an assessment of information recall and recognition skills, was found to improve understanding, especially among the participants with depression and the control group (*ibid.*). What Stiles et al.'s study highlights is the need for supported decision-making involving human interaction to be responsive to the particular experiences of cognitively vulnerable individuals if it is to fulfil its aim. That certain practices of supported decision-making are found to be unsuccessful with some cognitively vulnerable individuals is inevitable and should not dissuade us from searching for the most suitable approach.

[658] Benaroyo and Widdershoven (2004: 298). See also Welie and Berghmans (2006: 78).

[659] Bursztajn et al. (1991: 387) claim that competence assessment should be regarded as a "dynamic process" as opposed to simply a test in order for it to function as part of a wider process to develop autonomy. They do not discuss whether dynamism is a conceptual feature of competence, however.

Older Children and Adolescents

Since the mid-1980s, a number of studies have discussed the competence of children to consent to medical procedures, many of them in relation to competence to consent to treatment rather than research.[660] The majority of empirical studies that have considered the decisional competence of young people to consent to research have focused upon understanding, voluntariness and ratiocination as opposed to other elements of competence, such as appreciation or communication.[661]

Around the age of 9–11 years, children's understanding of disease, health and medicine appears to become more complex.[662] A seminal piece of empirical research from the early 1980s conducted by Lois Weithorn and Susan Campbell suggested that by the age of fourteen, adolescents have "generally equivalent" decisional competence to make informed consent treatment decisions as adults of twenty-one years.[663] At fourteen, the young people involved in the study were capable of understanding the nature and purpose of different treatment options, and most importantly, accepting the consequences for themselves of giving permission for health professionals to go ahead with a proposed form of treatment. More recently, Toner and Schwartz have applied this insight to the context of research,[664] arguing for a variant of the 'rule of sevens' where there would be a rebuttable presumption in favour of decisional competence to consent to biomedical research above the age of fourteen.[665] This resembles an earlier argument by Nicholson, although it differs in the respect that Nicholson would allow parental refusal to override an adolescent's consent to non-therapeutic research above the age of fourteen.[666]

It is important that in generating hypotheses about the decisional competence of adolescents, we do not lose sight of the social context in which decision-making takes place.[667] Socio-economic, educational and ethnic backgrounds have a significant role in the way a young person's decisional competence is developed and expressed.[668] So too is how parents, teachers and other authority

---

[660] Alderson (1993, esp. Chapter 9), Alderson and Montgomery (1996) and Fundudis (2003) focus on treatment whereas Keith-Spiegel (1983) and Nicholson (1986, Chapter 7) discuss research. Miller et al. (2004) provide a helpful overview of empirical studies on children's competence to consent.

[661] Miller et al. (2004: 260).

[662] Ondrusek (1998: 163), discussing Bauer (1990).

[663] Weithorn and Campbell (1982: 1595).

[664] Toner and Schwartz (2003).

[665] Toner and Schwartz (2003: 39). The rule of sevens is a concept derived from old English common law, where children were held to have no capacity until the age of seven, to be presumed to lack capacity between the ages of seven and fourteen and to possess capacity over the age of fourteen (*ibid.*).

[666] Nicholson (1986: 151).

[667] Alderson (1992: 119).

[668] Alderson (1992: 122), Weithorn and Schreber (1994: 168).

figures shape the decisional making context by circumscribing the child's power to decide.[669] Moreover, adolescents could also be though of as "doubly vulnerable" research participants, insofar as they often lack legal capacity to consent to research on their own behalf and are at greater risk of being manipulated into participation, due to their greater reliance on adults in the way information is presented to them.[670] Kipnis, alternatively, is sceptical of the claim that some types of vulnerability only apply to the child research participant and argues that they are evident in all research populations, albeit to different degrees.[671]

When research is conducted in a hospital environment, it may be more difficult for the young people concerned to understand that the hospitalization is connected with the research, rather than for some therapeutic purpose. In a study conducted by Schwartz into the child's understanding of involvement in a growth hormone study, less than a third of the cohort of 11–18 year olds (6 out of 19) appeared to understand that their hospitalization was due to their participation in the research.[672] Yet in another study, a considerable proportion of twelfth-grade college students (62%) appeared to understand the concept of 'vested interests' in research.[673] On the face of it, these findings represent inconsistent evidence about the decisional competence of young people. One explanation, which offsets this inconsistency, might be that the notion of self-interest is easier to grasp when it is practised in an environment not associated with beneficence.

Suspicions that age might have a bearing on decisional competence, along with the gender and background of the participants and their previous experience of research, are not universally borne out by the findings. In another study published by Tait et al., the authors could find no connection between understanding and gender, ethnic background, type of research or previous participation in research. Most surprisingly perhaps was that understanding did not differ with time taken to disclose information or to make a decision, although it still could have differed in relation to the presentational format used, which the authors did not investigate. Less surprisingly, there was a positive correlation between understanding and older age.[674] There is also some evidence from another study to suggest that adolescent females display a higher level of decisional competence than males of the same age range.[675] Developmental differences in the adolescent brain associated with gender may explain this.[676]

---

[669] Ashcroft et al. (2003: 17).

[670] See the discussion in Kopelman (2004), from which the term 'doubly vulnerable' is taken.

[671] Kipnis (2003: 119).

[672] Schwartz (1972) discussed in Miller et al. (2004: 267).

[673] Lewis (1981) reported in Miller et al. (2004: 273).

[674] Tait et al. (2003) discussed in Miller et al. (2004: 267).

[675] Mulvey and Peeples (1996) reported in Miller et al. (2004: 275).

[676] For a discussion see Reiss et al. (1996).

Ondrusek et al. conducted a significant empirical study in the late 1990s with eighteen children and young adults aged 5–18 years of age. The purpose was to ascertain the quality of consent amongst children who had already agreed to take part in a study relating to nutrition. All participants above the age of nine (with one exception) identified one strand of the purpose of the study; only four of the eighteen participants identified both purposes of the study (calorie measurement and improved medical knowledge).[677] It is interesting, however, that the study seemed not to utilise attempts at supported decision-making – the children were either read the form describing the study (if they were under thirteen) or given it to read for themselves (if they were above thirteen). The participants were only provided with further information if they or their parents asked questions.

Amongst the entire age group, there was a feeling that those involved with the research (perceived as authority figures) would be personally unhappy or disappointed if the participant exercised their right to withdraw. The authors believed this would manifest itself as a reluctance to withdraw should the child or young person reconsider their wish to be involved in the research.[678] This misperception clearly compromises the appreciation strand of competence, although one might well speculate that it is not necessarily confined to children – some healthy adult volunteers may harbour similar concerns. Perhaps most concerning is the statistic that only three out of the eighteen volunteers gave altruistic reasons as their motive for participating in research.[679] Of the four participants who were over seventeen years of age, only one gave such an altruistic reason.[680]

The research of Ondrusek et al. sounds a note of caution, particularly in respect of the consent of younger children.[681] However, in the case of older children it is less conclusive, particularly given that the authors did not investigate the impact that supportive measures might have on improving understanding. Given the type of comprehension problems displayed, supportive measures would seem best employed in assisting the young person to understand the methodological vocabulary of the research. Foster has rightly argued that for a researcher to consider a young person decisionally competent to consent to research, the young person would need to be able to understand methodological concepts involved in biomedical research such as randomisation and blind allocation of treatment.[682] It is also equally desirable for the young person to have a grasp of the normative concepts of

[677] Ondrusek et al. (1998: 160).

[678] Ondrusek et al. (1998: 163).

[679] Ondrusek et al. (1998: 164).

[680] Ondrusek et al. (1998: 164).

[681] Ondrusek et al. (1998). This was the group to whom they reserved their primary conclusion, which was to raise the age at which assent is sought from children in clinical research from seven to nine years of age.

[682] Foster (2001: 58).

altruism or supererogation, even if she could not articulate them in those terms.[683] An attempt at supported decision-making should seek to promote this understanding, and we should be inclined to treat an inability to grasp salient concepts involved in or raised by the research as evidence against competence.

That many adolescents have developing or under-developed emotional abilities is another factor that is likely to affect decisional competence to consent to research. Tan and Fegert argue, in relation to treatment, that there may be a discrepancy between the presence of intellectual requirements for understanding and their emotional correlates, which may not develop until some time later.[684] In addition, the experience of mental disorder or family problems might significantly delay the emergence of these emotional abilities.[685] Tan and Fegert suggest that this disjunction of intellectual reasoning and emotional insight may impinge upon decisional competence, particularly for more important treatment decisions.[686] If we attach greater salience to the emotions than have standard non-cognitive accounts, then it is questionable whether not only understanding but also appreciation can genuinely exist in such circumstances. Louis Charland, for instance, persuasively argues that to claim decisional competence can exist in the absence of an observable emotional repertoire rests upon a misconstrual of the nature of practical reasoning.[687] Indeed, if we accept that emotions are an expression of cognition rather than its antithesis, then any assessment of decisional competence must also account for the "positive contributing factor" of emotion.[688]

Overall, there is a dearth of empirical studies examining the decisional competence of young people to consent to research, which undermines our ability to draw definitive conclusions. From the evidence that is available, following mid-adolescence, there is a difference between instances of decisional competence in relation to treatment and instances of decisional competence in relation to research participation. However, it is entirely plausible to suggest that some

[683] This touches on the separate question of whether the young person has an obligation to participate in biomedical research. Although we need not venture too far into discussions of obligation here, it is sufficient to say that the political and economic context in which research place takes influences the force of any such obligation. The practice of biomedical research in a global deregulated market economy gives rise to concerns about exploitation that would not exist in an ideal-type Gewirthian community of rights. I have some sympathy with the view taken by Harris and Holm (2003) and Harris (2005a) that there is a moral obligation for adults and children to participate in biomedical research in particular contexts, and believe that this is broadly compatible with the PGC. However, the exploitative potential of global capitalism considerably undermines this obligation, especially where the benefits of medical research are liable to be used for private profit rather than shared for community benefit. For a discussion of the problematic ramifications of corporate culture for biomedical research, see Annas (1999) and Krimsky (2003).

[684] Tan and Fegert (2004: 291). See also McCabe (1996) for a discussion of developmental issues.

[685] Tan and Fegert (2004: 291).

[686] Tan and Fegert (2004: 291).

[687] Charland (1998a).

[688] Charland (1998b: 360).

adolescents – even if only a minority – have the decisional competence required to offer first-person informed consent to biomedical research. This is no different from recognising that some adolescents are competent to go scuba diving or to sit an advanced level mathematics exam several years early. That many adolescents are given very important responsibilities through acting as babysitters or acting as carers for family members casts doubt on the claim that they should, *prima facie*, be considered decisionally incompetent to consent to research.[689]

Judging a young person's decisional competence to consent to biomedical research turns upon whether there is sufficient evidence to show that the young person is capable of understanding and appreciating the consequences for herself of participation. This requires us to consider the young person's emotional integrity as much as conventional 'intellectual' aspects of decision-making. Nevertheless, an awareness of emotional integrity alone will not necessarily help us reach a judgment. To create a suitably detailed account of the young person's decisional abilities, we must engage appropriate forms of supported decision-making. Only once we have exhausted the means of support available for the young person to develop decisional competence to no avail can PTDCJ lead us to the conclusion that a young person is, on the evidence, unable to consent or refuse research participation on her own behalf.

Adults with Intellectual Disabilities

Regrettably, there is a lack of studies which examine the competence of adults with intellectual disabilities to consent to biomedical research, and very few on consent to treatment.[690] This is clearly problematic from the perspective of seeking an empirically informed view on the decisional competence of this group to consent to research. However, the treatment-based studies offer some insight as to how we may seek to improve decisional competence in a research setting, especially where supportive measures are used. This is not to say we can pre-empt the outcome, but we have greater understanding of the function of the process.

Arscott et al. adapted a vignette-based competence assessment instrument devised in an earlier study by Morris et al. and used it to ascertain the decisional competence of forty participants with a learning disability.[691] Three vignettes were used, describing a restraint scenario, psychiatric intervention and medical intervention. Whilst only 12.5% of the participants were found to be competent to

---

[689] Koren et al. (1993: 147).

[690] To date, these studies include Morris et al. (1993), Arscott et al. (1999) and Wong et al. (2000). One study which did examine consent to a research project involving adults with learning disabilities (Cameron and Murphy, 2006) related to the effectiveness of visual communication techniques rather than biomedical research.

[691] Arscott et al. (1999), Morris et al. (1993).

consent to each of the three vignettes, 65% of the participants were found to have sufficient decisional competence on at least one vignette.[692] Of these, participants were found to be competent to consent to the vignette involving surgery more commonly than the psychiatric or restraint vignettes. This contrasts with the findings in Morris et al., who found that 40% of a cohort of fifteen participants with mild learning disabilities had decisional competence to consent to treatment. In a group with severe learning disabilities, this was as low as one participant out of fifteen.[693]

Arscott et al. accept that an improved performance on the vignette relating to psychiatric medical treatment may have been expected, given that all the participants in the study were taking medication at the time of involvement; however, such an improved performance did not materialize.[694] In their view, this supports the claim that individuals with learning disabilities in general lack sufficient understanding about their medication.[695] In spite of the encouraging finding about competence on one or more of the vignettes, a failure to understand in the medical context suggests that understanding in the context of biomedical research may be even more limited. On this basis, we should be cautious about reaching a judgment of decisional competence.

Wong et al. sought to improve decisional competence with twenty adults who had a learning disability.[696] Those who were found to lack competence were not approached, unless they demonstrated a willingness to participate.[697] The participants were asked to give their consent to a blood test, which they had been advised to take by their GP. The authors provided an uncomplicated printed explanation of the study for potential participants to read, followed by a verbal summary.[698] Two means of explanation were used – continuous explanation and systematic explanation of each stage. The authors also limited the response required from the potential participants to basic recognition and gestural acknowledgment.[699]

The authors assessed decisional competence using a semi-structured interview which drew on the content of the MacArthur Competence Assessment Tool for Treatment (MacCAT-T). They found that when the information is presented in a more simplified way, decisional abilities increase.[700] Encouragingly, 65% of the participants with learning disabilities were found to have decisional competence.[701]

---

692 Arscott et al. (1999: 1371).

693 Morris et al. (1993), discussed in Arscott et al. (1999: 1368).

694 Arscott et al. (1999: 1372).

695 Arscott et al. (1999: 1373).

696 Wong et al. (2000). The authors also examined similar sized groups of participants with schizophrenia and dementia. This research is also presented as part of the discussion in Gunn et al. (1999).

697 Wong et al. (2000: 297).

698 Wong et al. (2000: 297).

699 Wong et al. (2000: 304).

700 Wong et al. (2000: 304).

701 Wong et al. (2000: 302).

This contrasted with 100% of the healthy control group.[702] However, the learning disability group encountered particular problems in distinguishing between the risk involved in the procedure and the risks of refusing the treatment.[703] Whilst risk to the participant of refusal would not arise in the case of research without a therapeutic element, this failure of appreciation does not augur well for the development of decisional abilities within a research context.

Taken together, both studies suggest that the scope for developing decisional competence in individuals with intellectual disabilities is limited, but not impossible. What transpires most clearly is the benefit of effective transmission of information. This need not be verbal communication, but constitute the most appropriate form for the particular participant. Thus, we can understand attempts to develop decisional competence in individuals with intellectual disabilities as a paradigm case of 'participant centred' consent seeking. This ethos is a prerequisite to PTDCJ, but we need to be clearer about how it could be sustained in the research context. As illuminating as these two studies are, there remains a great need for empirical research into the decisional competence of individuals with intellectual disabilities to consent to biomedical research.

Adults with Depression

The relationship between depression and competence to consent in healthcare contexts has been more widely researched than intellectual disability. This may be because, as Reynolds has acknowledged, psychiatric research could find itself in need of justification without empirical research into the decisional competence of potential participants with a mental disorder.[704] Yet depression has not received the level of attention that schizophrenia and dementia has, arguably because it is not regarded as a condition which would engage doubts about decisional competence, at least not in milder forms.[705] Carl Elliott explains the reasons for this:

> Most accounts of competence focus on intellectual capacities and abilities to reason, and depression is primarily a disorder of mood. According to conventional thinking, depression is primarily about despair, guilt and a loss of motivation, while competence is about the ability to reason, to deliberate, to compare and to evaluate. Often these latter abilities are ones that depression leaves intact.[706]

One of the challenges for empirical research into the decisional competence of individuals with depression is to shed light upon whether and to what extent the sufferer's disordered mood impacts upon these intellectual abilities of competence.

---

[702] Wong et al. (2000: 302).

[703] Wong et al. (2000: 303).

[704] Reynolds (2002: 117).

[705] Elliott (1997: 113).

[706] Elliott (1997: 113).

This bears upon, for example, whether we can say that a person with depression is "in equipoise" with respect to the decision as to participation or whether in reality she does not care either way.[707] The conclusions open to us may be somewhat tentative, however, as studies into the ability of depressed patients to consent to treatment outnumber those that pertain to their involvement in research, and a number of research-orientated studies focus on competence to consent to ECT, psychiatric and psychotherapeutic research rather than biomedical research *per se*.[708]

Following a study published by Grisso and Appelbaum in the mid-1990s in which 76% of patients with major depression were found to perform adequately across all aspects of decision-making in relation to treatment,[709] Appelbaum et al. sought to investigate the decisional abilities of depressed individuals in relation to research.[710] They used the MacCAT-CR to investigate the decisional competence of twenty-six women suffering from major depression to consent to a psychotherapeutic research study.[711] The authors' principal aim was to investigate the relationship between the severity of the symptoms and decisional abilities over time.[712] They found that the women who were not receiving treatment in hospital and whose depression was judged to be at least moderate "performed quite well" when their decisional competence to consent to research was assessed.[713] Although a significant number of participants encountered difficulties on the appreciation measure,[714] only a handful of participants displayed deficiencies in their understanding, reasoning or appreciation that would, in the view of the authors, bring their decisional competence into doubt.[715]

This is a surprising discovery, and one that the authors recognise is counter-intuitive.[716] That the depression was found not to affect decisional competence in the way it was expected requires corroboration by other studies in order for the findings to be understood as representative.[717] This is particularly important given the small research population from which the data was drawn.[718] However, we can reasonably conclude from this study that whilst the experience of depression does tend to adversely affect the decisional competence of the individual concerned, it is not an absolute quality and instead varies in degree from a mild effect compatible

[707] Charland (1998b: 373, n. 4).

[708] Such as the studies reported in Roth et al. (1982) and Appelbaum et al. (1999). An exception to this is Misra and Ganzini (2004).

[709] Grisso and Appelbaum (1995: 171).

[710] Appelbaum et al. (1999).

[711] Appelbaum et al. (1999: 1381).

[712] Appelbaum et al. (1999: 1381).

[713] Appelbaum et al. (1999: 1383).

[714] Appelbaum et al. (1999: 1383).

[715] Appelbaum et al. (1999: 1383).

[716] Appelbaum et al. (1999: 1383).

[717] Appelbaum et al. (1999: 1383).

[718] Appelbaum et al. (1999: 1383).

with continuance of many decisional competences to a more severe effect which erodes these considerably.

By contrast, Carl Elliott takes a more sceptical view. He argues that there is a significant risk that the experience of severe depression will erode decisional competence to consent to research altogether.[719] Elliott does not found his argument on empirical research of his own, but draws upon insights gained from previous studies of competence to consent to treatment. Considering the interaction between the effect of the mood disorder upon the appreciation of risk, Elliott claims:

> [I]f a person is depressed, he or she may be aware that a protocol carries risks, but simply not care about those risks. This sort of intellectual impairment can be as important a part of patient competence as the more detached, intellectual understanding that most accounts of competence emphasize.[720]

In support of this, Elliott refers to a study undertaken by Roth et al. in which a depressed woman who was otherwise deemed to be competent to understand the electroconvulsive therapy (ECT) she was offered, commented on the 1 in 3,000 chance of death associated with procedure by saying "I hope I am the one."[721] For Elliott, a failure to care about the risk involved, or to perceive the risk as a benefit, is tantamount to a failure of appreciation.[722] This is particularly relevant in research, since regard for one's own well-being is a basic precondition of an adequate assessment of research risks.[723] A choice arrived at in the absence of such self-regard lacks authenticity, as we cannot hold the decision-maker to account for it.[724] Since the lack of accountability flows from the affective dysfunction that impairs the decisional competence rather than the level of risk presented by the choice, Elliott's argument may apply beyond the particular choice expressed to any other choice that the decision in question allows.

The approach under PTDCJ differs slightly. We would require more detail about the nature of this woman's wishes before reaching a judgment of decisional competence in this treatment scenario. If her wish for her life to end is reasoned, long-standing and not the main reason she was consenting to the ECT, then we cannot rule out a judgment of decisional competence. However, if the wish is the main reason she is consenting, irrespective of its authenticity, then we might be inclined to understand it as an impediment to her appreciation of the ECT. Her wish is likely to be inauthentic if it fluctuates with the symptoms of her depression, abates with medication or is linked with the depression in some other way. In both cases, we should be inclined towards a judgment of decisional incompetence. By the same token, supposing the woman consented to take part in medical research because she

---

719 Elliott (1997: 113).

720 Elliott (1997: 113).

721 Elliott (1997: 114), discussing Roth et al. (1977: 282).

722 Elliott (1997: 114). Elliott draws on Bursztajn et al. (1991) in support of this.

723 Elliott (1997: 115).

724 Elliott (1997: 115).

wishes to increase her chance of death or serious injury, this would also constitute a manifest failure of appreciation. The expression of this wish in the context of research should incline one toward a judgment of decisional incompetence.

Neither Appelbaum et al. nor Elliott specifically consider the role of supported decision-making in enhancing decisional competence. A study that examined the impact of educative provisions on the decisional competence of depressed persons in relation to treatment therefore warrants discussion here. Lapid et al. investigated the decisional competence of elderly depressed persons to consent to ECT.[725] Participants were allocated into groups based on age (above sixty-five years and below sixty-five years). No participant had been judged incompetent to give consent on a previous occasion.[726] The authors assessed the participants' decisional competence twice; once at the start of the research and again after the educative intervention had been provided. All participants were shown a videotape explaining the procedure, and following randomization, some participants received a 30-minute session with a psychiatrist to discuss frequently raised concerns about the procedure.[727]

The authors found that the intervention of the psychiatrist increased understanding, reasoning and choice (understood as measure on the MacCAT-T).[728] As a group, the depressed elderly persons were found to have sufficient decisional competence to consent for themselves, and had a higher level of decisional competence than the younger group, although the authors could not be certain whether this was connected with the level of severe depression being lower in the elderly group than in the younger group.[729] The authors emphasised that these findings illustrate the value of employing appropriate educative interventions in other contexts involving a process of seeking informed consent.[730]

Ultimately, these findings support the assertion that depressed persons may have the ability for decisional competence to consent to research, but that this depends upon the nature and severity of the affective dysfunction and the effect on understanding and appreciation. As is the case with adults with intellectual disabilities, there is a greater need for empirical studies here, especially to ascertain how widespread are the type of findings reported by Appelbaum et al., and to what extent they would arise in the context of biomedical rather than psychotherapeutic research.

One possible direction for further research into the decisional competence of depressed persons is to focus on preferences evinced by the depressed person before the onset of their condition (if known) and to contrast them with those they display since developing the condition.[731] This would offer insight into the authenticity of preferences, and inform a judgment of whether decisional competence is

[725] Lapid et al. (2004).

[726] Lapid et al. (2004: 43).

[727] Lapid et al. (2004: 43).

[728] Lapid et al. (2004: 46).

[729] Lapid et al. (2004: 46).

[730] Lapid et al. (2004: 46).

[731] Rudnick (2002: 153).

present or not. Such an approach appears promising. However, it still leaves open the question of whether we can distinguish between a case where the person has changed her preferences since the onset of depression for reasons that are competently held and a case where her preferences appear to have changed due to volitional impairments following from affective dysfunction.

When applying PTDCJ to distinguish between such cases, we should consider the extent to which the potential participant appears to value her well-being as an instrumental good, even if not as an end in itself.[732] Such a basic concern for well-being is a precondition for holding preferences about research, which meshes with Elliott's argument. Judging the decisional competence of a depressed individual to consent to research also requires us to consider carefully the volitional content of the preference. An apparent absence of volitional content, resulting from a diminished ability to understand and appreciate the circumstances of the decision should lead us to the view that decisional competence is absent, due to the more severe consequences for the agent's rights of a false positive rather than a false negative judgment of competence. Insofar as supportive decision-making practices allow a better insight into the volitional abilities of a depressed individual, we have good reason to use them in the competence assessment process.

Adults with Schizophrenia

Decisional competence in schizophrenia tends to be determined principally by psychiatric aspects of the condition, such as avolitional states and inappropriate affect, rather than those symptoms of schizophrenia that may suggest themselves as obvious candidates for decisional incompetence, such as hallucinations and delusions.[733] This means that certain determinants of decisional competence in schizophrenia will be shared more widely by other mental disorders.[734] However, studies into the decisional competence of schizophrenia sufferers are more numerous than the other cognitively vulnerable groups we have so far considered. The reasons for this are not clear, although it may be a consequence of the importance of schizophrenia sufferers' participation in research, combined with an increasing interest in 'empirical ethics'.[735] In time, one hopes this may extend to adults with depression and intellectual disabilities.

Some commentators have sought to question the over-riding emphasis upon competence. Candilis has argued that the focus should not be on competence alone; instead, empirical studies should consider the values which schizophrenia sufferers

---

[732] *RM*: 52, Gewirth (1982: 118). See also the discussion in Beyleveld (1991: 23–24, 28–30, 177–181).

[733] Moser et al. (2002: 1205).

[734] Moser et al. (2002: 1206).

[735] Dunn (2006). Dunn defines empirical ethics as where empirical methodologies are brought to bear on the discussion of ethically problematic issues (*ibid.*).

are bringing to bear on their decisions whether or not to participate in research.[736] The problem with such an approach is that it rests upon an erroneous separation of values and competence. If one includes the capacity to value within a definition of decisional competence (which PTDCJ presupposes), then any attempt to consider values as a discrete line of inquiry from competence could serve to unhelpfully bifurcate future research into these inter-related elements of decision-making.

There is some evidence to support the claim that sufferers of schizophrenia value the benefits of research into schizophrenia and believe it should continue to go ahead. In a study conducted by Roberts et al., the views of schizophrenia sufferers were sought surrounding research participation.[737] The responses offered were enlightened and sophisticated. A majority agreed that participation was important for altruistic reasons and for the advancement of scientific understanding into the condition.[738] There was also a sense in which it gave rise to hope that other sufferers may receive better treatment.[739] This endorses the view that a considerable number of schizophrenia sufferers can come to a decision about research participation based upon an appreciation of the research derived from stable set of values. However, Roberts et al. claim that it may give rise to the risk of a "motivational vulnerability", where schizophrenia sufferers take part in research that is not in their best interests as a patient.[740] It is not clear exactly what the authors mean by this. If a schizophrenia sufferer is judged to have decisional competence to consent within a particular research context, then it follows that she should have the psychological ability to evaluate the motivational pressures to which the context gives rise, and make a decision for herself based on her own value system.

The main empirical studies to have been published this decade have recognised the significance of supported decision-making practices in eliciting the decisional competence of schizophrenia suffers. The first of these to be reported was Carpenter et al.,[741] who assessed the decisional competence of thirty research participants with schizophrenia and twenty-four healthy comparison participants to consent to clinical research.[742] The authors measured the cognition of the participants with schizophrenia against those of the healthy volunteers, and measured the decisional competence of all using the MacCAT-CR.[743] The authors concluded that the schizophrenia sufferers did demonstrate significantly poorer performance in respect of decisional capacities relevant to consent to research.[744] However, this did not reflect a persistent inability to understand the information relevant to a research study:

---

[736] Candilis (2003: 79).

[737] Roberts et al. (2000).

[738] Roberts et al. (2000: 71–72).

[739] Roberts et al. (2000: 71).

[740] Roberts et al. (2000: 73).

[741] Carpenter et al. (2000).

[742] Carpenter et al. (2000: 534).

[743] Appelbaum and Grisso (2001).

[744] Carpenter et al. (2000: 536).

> When offered additional opportunities to learn the necessary data, most subjects who scored below an *a priori* cut-off were able to bring their scores into the range of a comparison group of people without schizophrenia. This suggests that people with severe forms of schizophrenia may be able to give informed consent for research, although a single session brief presentation of research procedures may not be sufficient. Rather, an informed consent process that engages potential subjects over time and is sensitive to the negative impact of cognitive impairment may be essential for adequate informed consent.[745]

Carpenter et al. rightly emphasise the place of education in developing competence to consent to medical research. It cannot be over-emphasized that many individuals will require knowledge about the nature and purpose of the research presented in ways that take more time, effort and imagination than standard consent processes allow, and this is no more relevant than in cases where the potential participants have cognitive vulnerability of some kind. As we have already seen, decisional competence is a dynamic quality capable of development and, in the case of some cognitively vulnerable individuals, subject to volatility due to their mental state. This view has an influential precedent in contemporary psychiatric literature, but one that exists within the context of competence to consent to treatment rather than to research.[746]

In a more recent paper, Palmer and Jeste report the findings of a similarly important study.[747] They surveyed two groups of schizophrenia sufferers, those above seventy years of age and those aged between forty and seventy who had been diagnosed with schizophrenia under the DSM-IV classification without an associated diagnosis of dementia.[748] All participants were taking antipsychotic medication at the time of the study. The authors used a slightly modified version of the MacCAT-CR to ascertain each participant's decisional competence.[749] In addition, the authors employed a series of neuropsychological tests, designed to measure verbal and perceptual skills, along with working memory, cognitive processing speed and verbal and visual learning abilities.[750]

The authors found that the most significant determinant of decisional competence in both groups was performance on the neuropsychological tests.[751] Severity of general psychopathology was seen as a variable which could undermine the understanding element of decisional capacity, rather than depressive symptoms.[752] Most participants in the study had their decisional competence assessed as outpatients and, like Appelbaum et al. above, the authors acknowledge that the results may have been different if the research had involved more acute inpatients.[753]

---

745 Carpenter et al. (2000: 536).

746 Appelbaum and Roth (1981).

747 Palmer and Jeste (2006).

748 Palmer and Jeste (2006: 99).

749 Palmer and Jeste (2006: 100).

750 Palmer and Jeste (2006: 101).

751 Palmer and Jeste (2006: 100).

752 Palmer and Jeste (2006: 102–103).

753 Palmer and Jeste (2006: 104).

Again, the authors found that decisional competence improved with reiterative disclosure. This led the authors to conclude:

> The presence of schizophrenia (or other psychiatric diagnosis) alone is an inappropriate basis for determining decisional incapacity ...some of the impaired understanding after an initial disclosure should not be viewed as a deficit in consent capacity, but rather as a deficit in the commonly employed consent process.[754]

These findings highlight the relevance of using supported decision-making techniques in research with older schizophrenia sufferers. This is particularly important, given that the ability to comprehend consent forms is a skill that often declines in old age.[755] Elsewhere, Jeste et al. claim, "capacity should not necessarily be viewed as an unmodifiable trait, even when those patients initially seem to lack adequate capacity".[756] In support of this claim, they refer to a study conducted by most of the authors in 2001, reported in the *American Journal of Psychiatry*.[757] This study considered the effects of a randomized process to compare the effects of "routine" and "enhanced" consent with two groups of participants aged between forty and eighty years of age. One group were outpatients suffering from chronic schizophrenia and related psychosis; the other was a healthy control group.[758]

The principal difference between the styles of consent seeking was manifest in the form of the presentation and the time devoted to it. "Routine" consent simply involved a researcher who read out a written consent form with occasional pauses for questions and opportunities for the participant to seek clarification. By contrast, "enhanced" consent involved a Microsoft PowerPoint presentation structured into bullet points featuring summary slides relating to core information about the proposed research study.[759] The participants answered a quiz following the consent process which tested their understanding of key points. This became easier were they to fail to answer questions correctly. The participants had up to three chances, the second and third of which the participant was permitted to refer to the information on the original consent form.[760]

The results were illuminating. Scores on the first round of the test were considerably greater for schizophrenia sufferers who had been allocated to the enhanced consent group as opposed to the routine consent group.[761] This persisted when analysing the results of the second round of the quiz for those participants who took it. Most striking is that schizophrenia sufferers who experienced the enhanced consent procedure performed at an equivalent level on the first round of the test as healthy volunteers who were allocated to the routine consent arm of the study.[762] As the

---

[754] Palmer and Jeste (2006: 104–105).

[755] Jeste et al. (2003: 69).

[756] Jeste et al. (2003: 72).

[757] Dunn et al. (2001).

[758] Dunn et al. (2001: 1911), Jeste et al. (2003: 72).

[759] Dunn et al. (2001: 1912), Jeste et al. (2003: 72).

[760] Dunn et al. (2001: 1912), Jeste et al. (2003: 72).

[761] Dunn et al. (2001: 1912), Jeste et al. (2003: 72).

[762] Dunn et al. (2001: 1912–1913), Jeste et al. (2003: 72).

authors acknowledge, these results indicate the value of supported decision-making practices in eliciting the decisional competence of mentally disordered participants. Moreover, these findings are not isolated. Wirshing et al. discovered similar results in a study a few years earlier with repeated use of an informed consent questionnaire with schizophrenia sufferers.[763]

Another recently published study sought to analyse the effects of supported decision-making with schizophrenia sufferers. Moser et al. conducted a study to ascertain whether adding less than 30 minutes to the consent-seeking process could enhance the decisional competence of schizophrenia sufferers for inpatient and outpatient research.[764] The authors invited sixty participants to become involved in a hypothetical study designed to trial a drug to promote cognitive lucidity. Of these, thirty were diagnosed schizophrenia sufferers and thirty were healthy volunteers.[765] The authors designed a series of computer presentation slides with one main point per slide, which the participants with schizophrenia viewed whilst one of the authors read the main point aloud.[766] The authors assessed the participants' understanding using the MacCAT-CR before and after the intervention.

Prior to the intervention, there was a pronounced difference between the schizophrenia participants and the healthy control group, particularly in relation to understanding and appreciation.[767] However, the authors reported that individuals suffering from schizophrenia demonstrated "relatively strong" decisional competence even without supportive intervention, although this was not as high as the control group.[768] Following the intervention, the schizophrenia sufferers were "not significantly different from the healthy comparison group in any aspect of decisional capacity", namely understanding, appreciation, reasoning and expressing a choice.[769] As the hypothetical research study presented to the schizophrenia sufferers was more complex than those presented in previous research studies, the results of the study underscore more clearly the notion that supportive interventions can help sufferers to understand sophisticated procedures such as randomization, placebo-control or double-blindness.[770] Moser et al.'s findings show that schizophrenia sufferers' understanding can be improved through appropriate interventions and illustrates that the baseline level of cognitive impairment among schizophrenia sufferers may not be as great as one would think.

Overall, these studies constitute a convincing evidence base for why we should at least consider schizophrenia sufferers to have a developable decisional competence to offer consent to medical research, provided that disclosure of relevant

---

[763] Wirshing et al. (1998).
[764] Moser et al. (2006).
[765] Moser et al. (2006: 117).
[766] Moser et al. (2006: 117).
[767] Moser et al. (2006: 118).
[768] Moser et al. (2006: 118).
[769] Moser et al. (2006: 116).
[770] Moser et al. (2006: 119).

information is provided in a form that is conducive for understanding and appreciation. Of course, there remains a risk that participation authorised on the basis of first-person consent could be 'engineered' through processes that pitch the requirement for understanding and appreciation too low. Nevertheless, this represents an argument for greater scrutiny of supported decision-making practices by ethical review bodies, such as IRBs and RECs, rather than an argument against supported decision-making itself. Where we have a cognitively vulnerable group, such as schizophrenia sufferers, whose decisional competence to consent to research could be determined altogether by the type of consent process used, there is an imperative under PTDCJ to implement supportive decision-making practices that are conducive to preserving decisional competence as far as possible.

Adults with Alzheimer's Disease and Other Dementias

Changes in the way that dementia is perceived, both philosophically and clinically, has informed a view that dementia sufferers can and should take a more active role in research relating to their condition.[771] This is associated with wider discourses of empowerment in dementia.[772] On this view, the question of the decisional competence of dementia sufferers is left open, to be judged in relation to individual sufferers and in relation to the context of particular types of research. This progressive view is not universally shared, however – some commentators appear to start from the premise that dementia sufferers lack decisional competence altogether, and focus instead on the role of surrogate decision-making.[773]

This should not imply that surrogate decision-making is never an appropriate first response to dementia sufferers. A principal effect of dementia is impairment of the comprehension abilities required for decision-making,[774] and a recent longitudinal study has shown that this deteriorates sharply over two years from the initial stage.[775] Most persons with advanced dementia will be incapable of making decisions about research participation for themselves along with similarly sophisticated medical scenarios.[776] There is also evidence that relatively mild Alzheimer's disease can reduce decisional competence considerably in some cases.[777] However, the

[771] Bartlett and Martin (2002: 50).

[772] See Kitwood (1997) for a well-argued account.

[773] Yarborough (2002), for example, appears to do this.

[774] Buckles et al. (2003: 1662).

[775] Huthwaite et al. (2006).

[776] Kim et al. (2002a: 159). According to the studies which Kim et al. considered in their literature review, the proportion of dementia sufferers found to be decisionally incompetent ranged from 44% to 69% (*ibid.*). These studies were conducted largely in relation to competence to consent to medical treatment, however, and not research.

[777] Kim et al. (2001: 716).

onset of decisional incompetence in dementia is usually not a sudden but a gradual process in which the sufferer loses the ability to make particular decisions whilst retaining the ability to make others.[778] It is not fanciful to suppose, therefore, that some sufferers in the early stages of the condition may retain the decisional abilities to decide for themselves about research participation.

As is the case with three of the other four cognitively vulnerable groups considered in this book, insufficient empirical research has been conducted into the competence of persons with Alzheimer's disease or other dementias to consent to research,[779] but over the last few years more studies have begun to emerge.[780] In a literature review of efforts to assess decision-making ability in cognitive impaired (but not decisionally incompetent) elderly individuals, Kim et al. examined neuropsychological correlates of decisional impairment in Alzheimer's disease and other dementias.[781] Following the work of Marson et al.,[782] they suggest that tests of conceptualisation, word fluency and analysis of error behaviours can illuminate performance in executive functioning and assist in the identification of decisional abilities.[783] Kim et al. support the idea of a structured competence assessment test and entertain the possibility of a "consensus based criterion threshold" for measuring competence in relation to specific research studies,[784] although this would require agreement between clinicians and psychiatrists as to how to pitch the threshold, which may prove difficult to reach.

The authors claim that there is limited data to support the idea that educative interventions may improve scores on competence assessment procedures, but that the extent to which a dementia sufferer benefits will depend upon the severity of their cognitive impairment.[785] They conclude that supported decision-making may be possible until such time when severe impairment to executive functioning occurs.[786] This dovetails with Marson's findings that the pathogenesis of mild and moderate Alzheimer's disease leaves the communicative aspect of competence preserved.[787] Kim et al. note:

> by enhancing short-term memory (or perhaps bypassing the deficit by use of written memory aids in the consent process) in persons with mild dementia ... it may be possible to support the autonomy of those who may be seen as incapable by clinician-evaluators overly focused on the patients' memory deficits. The concept of comprehension is larger than mere free recall. In normal elderly subjects, the assessment of comprehension that relies on free recall underestimates the degree of actual comprehension. Thus, we may need to train

[778] Marson et al. (1994, 1997).
[779] Kim et al. (2002b: 799).
[780] Kim et al. (2002a: 151).
[781] Kim et al. (2002a: 159).
[782] Marson et al. (1997).
[783] Kim et al. (2002a: 159).
[784] Kim et al. (2002a: 161).
[785] Kim et al. (2002a: 162–163).
[786] Kim et al. (2002a: 163).
[787] Marson (2001: 281).

> competency evaluators to appreciate, and to help the patients make best use of, the intact strengths of those suffering from dementia.[788]

The espousal of supported decision-making is encouraging and represents an approach consistent with PTDCJ. However, it would be helpful to have more information on the experience of supported decision-making in real biomedical research contexts. One of the limitations of the research that Kim et al. examined is that most of the studies concerned hypothetical rather than actual decision-making scenarios.[789] Another limitation, from the perspective of gaining an insight into competence to consent to biomedical research, is that only a small proportion of the studies consulted (four out of thirty-two) considered decisional competence in this context. Most studies considered competence in relation to some other decision-making scenario, such as consent to treatment or competence to create an advance directive.[790] This indicates the need for further empirical work to validate existing findings in the area of competence to consent to biomedical research.

Other studies have investigated the extent to which Alzheimer's disease undermines comprehension of consent documents. Buckles et al. conducted a study in which 250 dementia sufferers with very mild, mild or moderate dementia were given a questionnaire to test their understanding of a proposed minimal-risk research study, about which their consent was being sought.[791] Statements such as "Information from this research study will help people who have memory problems" and "You are here for a birthday party" had to be designated true or false.[792] The authors also surveyed a control group of 165 healthy volunteers and adjusted for sufferers' educational attainment.[793]

All respondents in the healthy control group offered correct answers for at least eight out of the ten questions. The same was true of participants with very mild dementia. Of the sufferers with mild dementia, 92% were able to do this.[794] This contrasted with the group of moderate dementia sufferers, of whom 67% scored eight out of ten or above.[795] The authors recommended that the time and effort invested in presenting information could improve understanding,[796] especially by "in depth interaction and repetition".[797] However, the authors conceded that as the study focused on comprehension, it could not provide an insight as to how dementia affects appreciation, choice and reasoning, nor how comprehension might be impeded in a higher risk research study.[798]

---

788 Kim et al. (2002a: 160).
789 Kim et al. (2002a: 163).
790 Kim et al. (2002a: 154).
791 Buckles et al. (2003).
792 Buckles et al. (2003: 1664).
793 Buckles et al. (2003: 1663).
794 Buckles et al. (2003: 1662).
795 Buckles et al. (2003: 1664).
796 Buckles et al. (2003: 1665–1666).
797 Buckles et al. (2003: 1665).
798 Buckles et al. (2003: 1666).

In order to gain insight into the consequences of using psychiatric judgment and normative judgment about competence, Kim et al. measured the decisional competence of thirty-seven individuals with mild to moderate Alzheimer's disease and fifteen healthy comparison subjects of a similar age to offer consent to a fictitious biomedical research project.[799] The psychiatric judgment (delivered by three psychiatrists) was informed by an assessment of competence to consent using the MacCAT-CR. The psychiatrists assessed the decisional competence of both the Alzheimer's disease and comparison group. Although the members of the comparison group achieved highly on the competence assessment measure, the majority of the group with Alzheimer's disease demonstrated considerable decision-making impairment on at least one aspect of decision-making ability (62% on the three-psychiatrist criterion, 84% on the normative criterion).[800] Of the thirty-seven Alzheimer's disease participants involved, the authors judged ten decisionally competent to take part in the hypothetical clinical trial.[801] Whilst the primary purpose of this study was to compare the assessment of competence using these two criteria, it nonetheless offers a revealing insight into the decisional abilities of people with mild to moderate dementia.

The following year, Kim et al. published a study into the research participation preferences of Alzheimer's sufferers.[802] The objective of this was twofold. First, the authors intended to compare the willingness of Alzheimer's sufferers to participate in research protocols when compared with healthy elderly volunteers. Second, the authors wished to examine how degrees of decisional impairment in Alzheimer's sufferers affect their willingness to participate.[803] The authors selected a group of thirty-four sufferers with mild to moderate Alzheimer's disease and compared them with a control group of fourteen healthy elderly volunteers.[804] The decisional competence of the participants was sought in relation to four hypothetical research vignettes ranging from the mundane (withdrawal of blood) to the serious (brain surgery).[805]

The authors measured decisional competence according to the MMSE and the MacCAT-CR. The results of the study showed that, for three out of the four research protocols, there was no difference between the willingness of the Alzheimer's sufferers to participate and those of the control group of healthy volunteers.[806] Common to both groups was an inverse relationship between risk and willingness to participate.[807] The disparity between the two groups occurred in the context of the drugs trial, where the healthy volunteers showed less of a risk aversion to participation.[808]

---

[799] Kim et al. (2001).

[800] Kim et al. (2001: 715).

[801] Kim et al. (2001: 715).

[802] Kim et al. (2002b).

[803] Kim et al. (2002b: 797).

[804] Kim et al. (2002b: 797).

[805] Kim et al. (2002b: 798).

[806] Kim et al. (2002b: 799).

[807] Kim et al. (2002b: 799).

[808] Kim et al. (2002b: 799).

This corresponds to the authors' finding that deteriorating cognitive function induced by Alzheimer's disease appears to increase susceptibility to risk.[809]

We should not necessarily understand these results as an indicator of irrational deliberation amongst Alzheimer's sufferers, however. Although a sense of desperation which sometimes accompanies the initial stages of dementia is likely to erode in the later stages, a drive for self-preservation (e.g. not to have one's bodily integrity interfered with) survives serious dementia.[810] In this sense, there is a clear affinity between these findings and Jaworska's arguments about minimal valuing.[811] Severe Alzheimer's patients can still value aspects of what is at stake by research participation, even if they cannot understand the full implications of the risks or when they may be worth taking.

Not all commentators have implicitly accepted the premise that competence is vital for first-person consent to biomedical research with individuals suffering from dementia. David Guinn suggests that the emphasis on ascertaining competence in potential participants with dementia may be over-stated.[812] Instead, he proposes an 'affirmational' model of research with dementia suffers which would involve collaboration between the dementia sufferer and her legally authorized representative (LAR).[813] This derives from a notion of the research participant as "relational beings"[814] or an "individual-in-relationship",[815] which bears similarities with Hughes's concept of 'situated-embodied agency'.[816] The LAR derives their moral authority from being able to identify and further the dementia sufferer's values. Here, establishing whether a competent consent has been provided is less important than ensuring assent, as the risks that attach to a 'false positive' determination of competence would supposedly diminish due to the consent of the LAR, whose intimate knowledge of the dementia sufferer's values would in turn reduce the moral significance of doubt over the presence of decisional competence.[817]

Guinn's emphasis upon assisting the dementia sufferer's decision-making abilities through the involvement of an LAR is, in part, welcome. However, he appears to construe the moral authority of the LAR too strongly. As long as there remains a possibility that an individual possesses decisional competence, problems arise with the notion that someone with authority to decide for the individual could actually empower her in doing so. Until the possibility of decisional competence has dissipated, she should be encouraged to speak for herself. There is a moral as well

---

[809] Kim et al. (2002b: 799).

[810] Kim et al. (2002b: 801).

[811] Jaworska (1999, 2006), discussed in Chapter 5.

[812] Guinn (2002).

[813] Guinn envisages this ideally as the dementia sufferer's partner, a relative or a close friend who has a "loving and intimate knowledge of the subject's wishes or value system" (2002: 239).

[814] Guinn (2002: 230).

[815] Guinn (2002: 236).

[816] Hughes (2001), discussed in Chapter 5.

[817] Guinn (2002: 241).

as a practical difference between providing assistance in decision-making for those who do not have the means to make such decisions unaided and displacing the primacy of the person about whom the decision will affect by using a surrogate decision-maker. If we apply PTDCJ, then we should only involve an LAR when the available evidence suggests that competence is unlikely or where to presume competence would lead to the individual causing harm to themselves or others. The 'trigger point' for involvement of the LAR on Guinn's account would seem to occur much too early on.

In summary, Marson is right that more research needs to be undertaken with elderly persons not suffering from Alzheimer's disease in order that we can develop a richer appreciation of the effects of different types of neuro-degeneration on decisional competence.[818] The effects of prior participation in research have already been shown to strengthen the decision-making abilities of dementia sufferers[819] and ways to simulate this experience (e.g. through an interactive DVD presentation) might prove effective. In any event, it may be that elderly persons engage with decision-making processes differently from younger adults.[820]

Although there is clearly still work to do, the evidence presented here cannot support a complete prohibition on seeking first person consent from Alzheimer's disease and other dementia sufferers for participation in biomedical research. Given the wide variation of decisional competence that exists between such individuals, it would be more appropriate to proceed upon a case-by-case basis, which recognises that we can only judge Alzheimer's and other dementia sufferers to have lost particular decisional competences when they can no longer display them. This is consistent with PTDCJ in that we should look only to the decisional abilities of the individual in reaching a judgment of their competence to consent, and never allow the utility of the research to have a subliminal influence on the conclusions we reach.

Summary

From the evidence presented in this chapter, it is clear that there is no group about whom we can say that all members unequivocally lack the abilities to develop decisional competence to consent to research, as we could about very young children or the comatose. To a greater or lesser degree, all groups have displayed the potential for development of decisional competences. When applying the principle of PTDCJ, the level of certainty which a judgment of competence requires is determined by the impact of an ascription of decisional competence or incompetence on the agent's freedom and well-being and, where relevant, that of others. This is informed by the context of the decisional abilities we can reasonably expect someone in her position

[818] Marson (2001: 281).

[819] Rikkert et al. (1997: 274).

[820] Gatz (2006).

to display (if a presumptive judgment) or by a demonstration of decisional abilities (if a case-by-case judgment).[821] In this sense, the level of certainty PTDCJ requires differs from the standard proposed by Adamis et al., who argue that the certainty of a judgment of competence is proportional to the foreseeable risks involved in the activity.[822] Under PTDCJ, the certainty of the judgment is not proportional to the risks *tout court*, but proportional to the impact upon the generic rights of the decision maker and, where they are affected, those of other agents. So PTDCJ is risk-sensitive insofar as it is responsive to those risks that arise from the particular circumstances of a competence judgment which are relevant to and measured by the PGC.

When making a judgment of decisional competence under PTDCJ, we can broadly distinguish the empirical evidence for decisional competence into three categories: compelling, uncompelling and inconclusive. From the evidence presented in this chapter, the cognitively vulnerable individuals likely to fall into the 'compelling' category include mild to moderate schizophrenia sufferers, older adolescents and adults with mild forms of intellectual disability and dementia. The individuals likely to fall into the second category include adults with advanced dementia, major depression and severe intellectual disability. We should reserve the third category for those individuals for whom the existing empirical research has indicated little about their decisional competence, such as younger adolescents. The factors that place cognitively vulnerable individuals within these three categories include psychological maturity, ability to generate preferences in relation to specific decisional contexts, level of minimally adequate cognitive functioning (including emotional integrity) and the liability of decisional competence to recurrent fluctuation.

However, the limitations of the existing research attenuate the force of the conclusions we can draw. Notwithstanding the publication of new studies (most of which have emerged since the mid-1990s, with the exception of those relating to older children and adolescents), there remains a considerable need for more evidence on the decisional competence of cognitively vulnerable groups to consent to the specific context of biomedical research and methods to support their decision making. We should also note two further shortcomings. First is the absence of studies on supported decision-making with children, which is disappointing. Second, many of the empirical studies discussed do not undertake follow-up judgments of decisional competence to establish whether the grounds for making such judgments change in the medium to long term. We require more longitudinal studies, to ascertain in whom, how and why decisional competence fluctuates over time.[823]

---

[821] Judgments of decisional competence with cognitively vulnerable groups are, as we have seen, most likely to proceed from a case-by-case-assessment.

[822] Adamis et al. (2005: 138).

[823] Involving participants with fluctuating decisional competence in research would require involving an LAR or other surrogate decision-maker during periods of incompetence. The prospects for applying supported decision-making to advise individuals with fluctuating decisional competence on the consequences of research participation during a time where decisional competence is apparently present are encouraging, although the issues of balancing first-person consent with surrogate consent raise issues which lie outside of the scope of this book. On this, see further Stroup and Appelbaum (2003).

Judgments of decisional competence are, as we recognised at the beginning of the chapter, a synthesis of normative and empirical considerations. To make judgments of competence without extensive empirical knowledge reflects a significant deficit in our understanding of what it means to decide competently.[824] It is probable that poorly designed consent-seeking procedures in part explain the shortcomings in our knowledge of the decisional competence of cognitively vulnerable groups.[825] Even more than twenty-five years ago, Roth et al. voiced scepticism about the increasing length and complexity of consent forms, claiming that particularly difficult forms "may obstruct rather than improve the process of obtaining informed consent".[826] A failure to reflect on how to remedy this efficaciously could give rise to ill-conceived attempts to support the decision-making of cognitively vulnerable participants or pitching too low the level at which they need to demonstrate those decisional abilities required for a judgment of competence to be made.

To be successful, supported decision-making must be tailored to the particular needs of the potential participant, in the sense that it should be receptive to the nature and effect of her own cognitive vulnerabilities. Equally important is that supported decision-making practices recognise an appropriate level at which decisional competence should be demonstrated in particular contexts. This countervails against the possibility of inadvertently lowering the standard of decision-making competence required through using such methods. From the evidence presented in this chapter, the first priority of supported decision-making should be in maximizing understanding and appreciation, two factors that are often lacking in cognitively vulnerable participants. Provided we use this insight as a point of departure, we should not be discouraged from using imagination and creativity to devise suitable approaches. The methods discussed in this chapter already show evidence of this.

It is unlikely that there will emerge a panacean method for improving understanding in decisional competence.[827] Nevertheless, a common factor in the most effective forms of supported decision-making surveyed involve a significant element of human interaction, such as educators or counsellors spending more time with potential participants to answer their questions, to offer additional explanations and to provide reassurance. Even where technological aids are used (such as in the PowerPoint example), the human element is pivotal in their success. This would suggest that mere attempts to make an existing consent form easier to read or to give potential participants more time to digest information on their own will not be enough.

In conclusion, Guinn is right that individuals are relational beings, but their embeddedness within particular personal and social relationships does not displace the moral force of ensuring that as far as possible, they are permitted to make decisions for themselves about matters which affect their own physical and psychological well-being. Supported decision-making obviates the criticism that in seeking to

---

[824] Marson (2001: 280).

[825] Dunn and Jeste (2001: 604–605).

[826] Roth et al. (1982: 48).

[827] Dunn and Jeste (2001: 596).

preserve the autonomy of cognitively vulnerable persons, we are abandoning them to an autonomy that they do not possess. It should not be forgotten that a judgment of decisional incompetence not only denies an individual the authority to be the principal determinant of the outcome of a particular decision, it also can have serious psychological consequences for the individual in her broader life, affecting the way she perceives herself and how others respond to her.[828] We should only arrive at a judgment of decisional incompetence after the fullest possible assessment of competence involving a sustained and appropriate form of supported decision-making to attempt to elicit any latent decision-making abilities that she may have.

Scrutinizing extant empirical studies has illuminated our understanding of the decisional competences of cognitively vulnerable groups and methods to support their decision-making abilities. In taking stock of the insights discussed this chapter, we have drawn conclusions that relate both to the empirical evidence required for judgments of competence under PTDCJ and how practices of supported decision-making can be made to best fulfil their purpose. We now have a basis on which to consider how judgments of decisional competence to consent to biomedical research are articulated in a regulatory and legal context, and to investigate whether any parity exists between PTDCJ, codes of ethics and legal rules.

[828] Winick (1995: 8).

Chapter 7
Cognitive Vulnerability and Consent to Biomedical Research in England and Wales

Until recently, the status of first person contemporaneous legal competence (FPCLC) to consent to medical research in England and Wales was in considerable disarray. The absence of a legislative measure combined with ambiguous ethical guidelines and confusing terminology, created a situation fraught with uncertainty. To a considerable extent, this confusion persists despite efforts to systematize the law. Whilst competence and capacity are two of the most widely used concepts in relation to English medical law, they are also two of the least understood and most poorly employed. In part, this has arisen due to inadequate attention paid to assessments and judgments of decisional competence.

This confusion is starkly apparent amongst medical practitioners. In a survey published in 2002 regarding knowledge of consent and capacity in the medical profession,[829] only a small proportion were aware of the legal position relation to consent and research. Two out of sixteen junior doctors (13%), three out of ten non-career-grade doctors (30%), seven out of nineteen specialist registrars (37%), five out of fifteen consultants (33%) and four out thirty-five general practitioners (11%) gave correct answers.[830] For every grade, this was lower than the figures for consent to treatment. These alarming figures represent a dearth of knowledge on behalf of the medical profession about the current law. As such, it is difficult to imagine that these practitioners would be in a position to apply the existing law accurately, let alone evaluate its shortcomings.

The purpose of this chapter is to survey the legal position in England and Wales and selected ethical guidelines issued by professional bodies in the UK. It will also consider the significance of European regulation in generating norms of decisional competence judgment that have influenced the English legal position or may do so in time. Throughout, we will assess the prospects for compatibility between the regulatory position and precautionary task or decisional competence judgment (PTDCJ). In particular, we will consider the differences that exist between 'mental capacity' and 'legal capacity', the current legal status of decisional competence to

[829] Jackson and Warner (2002).

[830] Jackson and Warner (2002: 602).

P. Bielby, *Competence and Vulnerability in Biomedical Research*,

consent to medical research in England and Wales and ethical guidance on decisional competence to consent to medical research. As our focus is on FPCLC, we will not consider the position of decisionally incompetent individuals (frequently known as 'mentally incapacitated' or 'mentally incompetent' individuals) as they will in all probability lack FPCLC.

Three Approaches to Determining FPCLC

It is helpful to begin by drawing a distinction. A legal judgment of FPCLC can be made upon three possible grounds – on a functional, outcome or status based test.[831] Functional tests involve an assessment of the task or decisional competence of an individual in respect of a particular task or decision. A functional test of capacity involves ascertaining the task or decisional abilities of an individual in terms of the nature, purpose and effect of the activity in question at the time of the assessment. Functional tests of capacity therefore establish a clear link between decisional competence and FPCLC, so that the latter is typically explained mostly, if not wholly, in terms of the former.

Functional tests are particularly useful where an individual is associated with a cognitively vulnerable group who usually may not be considered to have task or decisional competence in respect of that activity. This is because such tests have the advantage of recognising both specific contexts in which the individual can be judged to have FPCLC and areas where she can be judged to lack capacity in this regard. Functional tests of capacity are analogous to demonstrable task or decisional competences given the demonstration of ability at the task or decision needed in order to convince the assessor of the presence of the ability and the specificity of the test to the task or decision involved.

Outcome tests of FPCLC are concerned with the content of the choice that the individual reaches. As such, they can be utilized only in decisional contexts. According to this test, a decision that is inconsistent with accepted values, clinical advice or with the judgment of the assessor represents sufficient grounds for declaring the individual to lack capacity in that respect. Outcome tests of FPCLC are closely related to asymmetrical or risk-relative models of competence we examined in Chapter 1. If applied, outcome tests potentially give rise to a situation whereby an individual is considered to have FPCLC if she chooses one way, but lacks FPCLC if she chooses another within the same decisional scenario. This approach to determining FPCLC is premised on values which tend towards paternalism rather than the promotion of individual autonomy and in extreme cases engineer conformity to received opinion. The nature of such a test in the context of determining consent to

[831] See Freedman (1981) for a theoretical discussion and Law Commission of England and Wales (1993, Part II, pp. 10–24) for a discussion in relation to English law.

participation in biomedical research could be open to exploitative misuse, as it is possible that the values of the competence assessor could surreptitiously determine the existence of FPCLC, disguised as an attempt to support decision-making.

Status tests confer FPCLC upon the grounds of the possession of a certain characteristic, such as age, achieved universally. It excludes all individuals who do not fall within this group as lacking this capacity. A status approach grounds the enfranchisement of individuals once they reach the age of eighteen years, and determines other decisional competences, such as contractual capacity and consent to sexual intercourse. Status approaches are less appropriate in healthcare contexts where the moral imperative of giving effect to decisional competence is high and the cost of assessing competence on a case-by-case basis is comparatively low. They also can give rise to, as the Law Commission noted, the presumption of an absence of FPCLC in *all* areas, where this is not the case.[832] However, status approaches may be necessary where adopting a functional approach would be administratively unworkable (such as a functional test for adolescents to attain the right to vote). In this sense, status approaches are analogous to presumed competences insofar as they require a universally attainable quality and the absence of a test of individual capability.

A functional test to determine FPCLC is most consistent with PTDCJ. This is because it ensures that the grounds upon which competence is judged are specific to the nature of the decision at hand, and do not account for the possession or absence of other task or decisional competences or instances of FPCLC that are irrelevant to the abilities being assessed. Status approaches tend to frustrate a PTDCJ approach to FPCLC as they rely upon blanket presumptions of decisional competence, unless the adoption of a status test is the only way FPCLC can be ascribed without undermining more important rights of other agents. Outcome tests accord an unwise amount of discretion to those responsible for assessing competence in ways that could allow inappropriate factors such as the values and wishes of the competence assessor to influence a determination of FPCLC.

The Meaning of FPCLC Consent in England and Wales

English law has originally formulated standards of FPCLC in relation to medical treatment rather than research. This is the case in the context of adults and children. These standards warrant consideration, as they inform the basis of the standard of FPCLC to consent to research, and would provide a legal background to any attempt in English law to develop specific principles in relation to making judgments of decisional competence.

[832] Law Commission of England and Wales (1995, para. 3.3).

Adults

Following the judgment of Lord Donaldson in *Re T,*[833] every adult in English law is presumed to have FPCLC consent to treatment, but this is a presumption that can be rebutted if the existence of capacity is brought into doubt.[834] The level of understanding required for a legally valid consent to treatment must be commensurate with the gravity of the decision to be taken, with more serious decisions requiring greater capacity.[835] This appears on the face of it to be an example of risk-relative competence. However, it is compatible with a decision-relative theory of competence, as the decisional abilities required are related to the complexity of the decision to be taken, not simply the choice made, and 'seriousness' is a measure of this complexity, rather than merely a measure of the risk.

The most authoritative English case law definition of FPCLC in adults is provided in *Re C (adult: refusal of medical treatment),* a case in which a schizophrenic patient in a psychiatric hospital refused consent for an operation to amputate his gangrenous leg.[836] Thorpe J held that in order to have FPCLC, to consent, the patient must understand "the nature, purpose and effect"[837] of the procedure. This involves the ability to comprehend and retain information about treatment, to believe this information and to weigh the information in order to evince a choice.[838] When we evaluate this test against non-legal criteria of decisional competence, it yields fundamental similarities with the conceptual definition of decision-relative decisional competence examined in Chapter 1.

More importantly, *Re C* firmly establishes that the quality of internal reasoning expressed by the decision-maker should solely determine FPCLC in adults, and not the view of the competence assessor about the reasonableness of the choice or the rationality of the reasoning process. This is again analogous to decision-relative competence, in that the consistency of the decision-maker's reasoning in terms of her own established values is interpreted as a indicator of decisional competence rather than how well the value judgments expressed in that reasoning process correspond with those health care professionals believe the patient ought to hold. This principle has been affirmed in England and Wales subsequently, in *Re MB (an adult: medical treatment),*[839] which concerned a pregnant woman refusing venepuncture for a caesarean section, and perhaps most notably in *Re B (adult: refusal of medical treatment),*[840] where a quadriplegic adult's refusal of life-sustaining treatment was granted.

---

[833] *Re T (adult: refusal of medical treatment)* [1992] 4 All ER 649.

[834] *Ibid.*, Lord Donaldson MR, p. 661.

[835] *Re MB (an adult: medical treatment)* (1997) 38 BMLR 175 (CA).

[836] [1994] 1 All ER 819.

[837] *Ibid.* Thorpe J, p. 824.

[838] *Ibid.* Thorpe J, p. 824.

[839] (1997) 38 BMLR 175 (CA).

[840] [2002] EWHC 429 (Fam).

Re C symbolizes a clear recognition of the acceptance of a functional test towards determining adult FPCLC in English law, as opposed to a status test. Accordingly, the test turns upon the decisional abilities of the individual and not upon any concomitant mental disorder or intellectual disability, provided that the existence of the disorder or disability does not affect the integrity of the decision-making. The existence of these important parallels between standards of decisional competence and tests of FPCLC in adults indicates how in English law, 'competence' is frequently understood as a synonym for 'capacity'. Indeed, it is true to say of the *Re C* test that it defines FPCLC wholly in terms of decisional competence.

The burden of proof in any dispute about FPCLC is placed upon those who are alleging that someone lacks FPCLC, or alternatively that someone who previously was judged to lack FPCLC now possesses FPCLC.[841] This meshes with the position under PTDCJ, as the basis for a judgment under precaution has to be that whatever is being alleged can be given a compelling explanation under conditions of uncertainty,[842] and that the consequences for the generic rights of the individual(s) concerned in believing one way are less harmful than the consequences in believing another. The standard of proof that a judgment of FPCLC requires is the civil law standard of the balance of probabilities rather than the higher criminal law standard of beyond all reasonable doubt.[843] Given the administrative difficulties that may follow from operationalizing a standard of beyond all reasonable doubt (that could, as we saw in Chapter 5, compromise the state's ability to fulfil its obligations to other agents), a balance of proof on the civil law standard appears appropriate from the perspective of PTDCJ.

Tan and McMillan argue that there is a disparity between the standard of legal capacity articulated in the *Re C* test and that which is embodied in the BMA and Law Society's guidance document, *Assessment of Mental Capacity*, the second edition of which was published in 2004.[844] This arises because the guidance couches FPCLC in terms of understanding and retention of information (concentrating on the first strand of the *Re C* test), while failing to elaborate on believing and weighing the information (the second and third elements of the *Re C* test).[845] The BMA and the Law Society also appear to add to the *Re C* test by including the freedom to make a choice, which is usually part of the definition of consent rather than competence or capacity.[846]

---

[841] Mental Capacity Act 2005, s. 1(2). See also the Mental Capacity Act 2005 Code of Practice, para. 4.10, p. 44 and the British Medical Association and the Law Society (2004: 28).

[842] Remembering that in matters of judging decisional competence, just as in matters of agency ascription, all that can be known with certainty is that I myself am an agent.

[843] Mental Capacity Act 2005, s. 2(4). See also the Mental Capacity Act 2005 Code of Practice, para. 4.10, p. 44 and the British Medical Association and the Law Society (2004: 29–30).

[844] Tan and McMillan (2004), British Medical Association and the Law Society (2004).

[845] Tan and McMillan (2004: 428).

[846] Tan and McMillan (2004: 427–428).

Tan and McMillan directed their criticism towards the first edition of the guidance, published in 1995. Although the second edition now includes reference to weighing the information in reaching a choice, and freedom of choice is now related explicitly to the validity of the consent, there is still no reference to believing the information.[847] It remains an important omission, since believing the information presented is as central to making a judgment of decisional competence as understanding the information – even if the decision-maker subsequently goes on to reject the importance of the information. Moreover, it is likely that healthcare professionals, one of the target audiences of the guidance, will come to interpret the *Re C* test in the more restrictive terms in which the BMA and the Law Society have construed it.[848]

The connection between FPCLC and decisional competence is strengthened the recent introduction of 'mental capacity' as a statutory legal concept by the Mental Capacity Act 2005 (hereafter MCA 2005).[849] Until recently, attempts have been made to maintain the separation of the concepts of legal and mental capacity.[850] Since then, there has been a retreat from this position to the point where the new statutory definition of legal capacity in English law is framed in terms of mental capacity. The Act provides a clear presumption in favour of FPCLC grounded in decisional competence[851] and applies to individuals above the age of sixteen years. Section 2(1–2) of the MCA 2005 provides:

> s. 2 People who lack capacity
>
> (1) For the purposes of this Act, a person lacks capacity in relation to a matter if at the material time he is unable to make a decision for himself in relation to the matter because of an impairment of, or a disturbance in the functioning of, the mind or brain.
> (2) It does not matter whether the impairment or disturbance is permanent or temporary.

Inability to make a decision is defined in s. 3(1)(a–d) of the Act as an inability to understand, retain, use and weigh the information relevant to the decision and an inability to communicate the decision (belief is not included as a criterion).[852] This tracks very closely the structure of decision-relative decisional competence to consent discussed in Chapter 1. By implication, mental incapacity in English law must be defined as the absence of those abilities. This posits a *prima facie*

[847] British Medical Association and the Law Society (2004: 120).

[848] Tan and McMillan (2004: 428).

[849] The MCA 2005 received Royal Assent on 7 April 2005 and came into force in two phases during 2007.

[850] Law Commission of England and Wales (1991: 19).

[851] MCA 2005, s. 1(2).

[852] A similar definition of capacity has been issued previously by the Department of Health in their *Good Practice in Consent Implementation Guide*: "A patient will lack capacity to consent to a particular intervention if he or she is ... unable to comprehend or retain information material to the decision, especially as to the consequences of having, or not having, the intervention in question, and/or ...unable to use and weigh this information in the decision-making process" (2001b: 46).

conceptual link between decisional competence and mental capacity and, by extension, to FPCLC as mental capacity provides the criteria for having FPCLC.

The Explanatory Notes to the MCA 2005 stipulate that the Act's definition of capacity "focuses on the particular time when a decision has to be made and on the particular matter to which the decision relates, not on any theoretical ability to make decisions generally".[853] In other words, a person's inability to manage their financial affairs, for example, should not influence judgments about their decisional competence in relation to consent to medical treatment. It follows from this that FPCLC strongly pertains to the ability of the individual to make a decision for him or herself that is decision-relative, and therefore constitutes a legal instantiation of decision-relative competence. Thus, FPCLC would appear to approximate to a statutory definition of mental capacity that is itself informed by decision-relative notions of decisional competence.

The MCA 2005 goes on to make provisions in relation to research with individuals who lack mental capacity that do not involve clinical trials, although these do not give rise to a separate test of FPCLC consent.[854] FPCLC to consent to medical research is therefore determined by the provisions of s. 3. Similarly, the Human Tissue Act 2004 makes provisions for "appropriate consent" but construes capacity in terms of its meaning under the MCA 2005.[855] We can gather from this that in terms of FPCLC to research, the test of capacity elaborated in s. 3 of the MCA 2005 along with the case law discussed above will remain the principal reference point.

The Medicines for Human Use (Clinical Trials) Regulations 2004 (hereafter 'the Clinical Trial Regulations') is a piece of delegated legislation (made under the authority of the relevant Secretary of State) which implements Directive 2001/20/EC into English law.[856] Schedule 1 Parts 1–5 of the Clinical Trial Regulations outline legal protections for adults, incapacitated adults and children involved in clinical trials. "Adult" is defined as an individual above the age of sixteen years.[857] Again, there is no explicit discussion of the principles underpinning decisional competence or FPCLC and no guidance on what approach should be taken if decisional competence is questionable or how to support decisional competence. However, the EU Directive, which was responsible for the British Government enacting the Clinical Trials Regulations, makes elliptical reference to FPCLC by presuming its absence in certain cognitively vulnerable participants. Paragraph 4 of Directive 2001/20/EC (hereafter 'the Clinical Trials Directive') discusses the participation of the mentally impaired:

---

853 Department of Constitutional Affairs and Department of Health (2005: 4).

854 MCA 2005, s. 30–34.

855 Human Tissue Act 2004, s. 1–3 and Explanatory Notes, Part 1, Section 6.

856 SI 2004/1031, s. 28–31 and Schedule 1. There is no general exclusion of tissue-based research in either the MCA 2005 or the Clinical Trial Regulations; however, the Human Tissue Act 2004 more comprehensively regulates the removal, storage and use of human tissue samples.

857 Clinical Trials Regulations 2004, s. 2(1).

> In the case of persons incapable of giving their consent, such as persons with dementia, psychiatric patients etc., inclusion in clinical trials in such cases should be on an even more restrictive basis. Medicinal products for trial may be administered to all such individuals only when there are grounds for assuming that the direct benefit to the patient outweighs the risks. Moreover, in such cases, the written consent of the patient's legal representative, given in cooperation with the treating doctor, is necessary before participation in any such clinical trial.[858]

The reference to the written consent of the participant's legal representative entrenches the assumption that all such participants would lack FPCLC and instead be entered into clinical trials though provisions for proxy consent.[859] This assumption links decisional incompetence to the presence of a mental disorder and returns us to the status based tests outlined above. This is a retrograde step, especially at a time when the general tenor of English case law and the MCA 2005 represents a significant departure from this trend.

In England, academic commentators and judges frequently use competence as a synonym for FPCLC.[860] Given the affinity between decisional competence and FPCLC, underscored through the advent of the MCA 2005, this is not surprising. Competence, however, is not a legal concept in English law.[861] This means that in current English law, FPCLC *can* be defined in terms of decisional competence *but this need not necessarily be the case*. It follows that it cannot be known with absolute certainty whether competence is being used to express a non-legal judgment about the decisional competence of the individual, or as a substitute for FPCLC. This militates against clarity and consistency in legal reasoning.

To illustrate this, consider the use of competence in the following excerpt from the judgment of Butler-Sloss LJ in *Re MB*[862]: "The graver the consequences of the decision, the commensurately greater the level of *competence* [emphasis added] is required to take the decision."[863] This claim, considered alone, is perfectly reasonable. However, in the next paragraph, Butler-Sloss LJ expounds the criteria for an individual to lack *capacity*.[864] It is unclear from this whether the use of 'competence'

---

[858] Directive 2001/20/EC, OJ L 121, 1.5.2001, p. 35.

[859] Additionally, the emphasis placed upon "direct benefit to the patient" means the Directive would permit only therapeutic research upon those participants incapable of consenting.

[860] See, for example, Brazier and Lobjoit (1991: 34–51), Kennedy and Grubb (2000, Chapter 5) and Mason and Laurie (2006, Chapter 10).

[861] This is antithetical to the position in the US, where 'capacity' denotes *mental* capacity only, and competence is a legal concept. See Berg et al. (1996: 348–349).

[862] (1997) 38 BMLR 175 (CA).

[863] Butler-Sloss LJ, p. 198.

[864] "A person lacks capacity if some impairment or disturbance of mental functioning renders the person unable to make a decision whether to consent to, or to refuse, treatment. That inability to make a decision will occur when: (a) the patient is unable to comprehend and retain the information which is material to the decision, especially as to the likely consequences of having, or not having, the treatment in question; (b) the patient is unable to use the information and weigh it in the balance as part of the process of arriving at the decision" *ibid.*, Butler-Sloss LJ.

is a reference to the psychological decisional abilities of MB, and the use of 'capacity' a reference to the legal factors that would render any individual to be without FPCLC to treatment.

A judge may, of course, disclaim any such difference between the two terms and rely upon the resemblance between non-legal criteria for decisional competence and FPCLC outlined above. However, closer analysis of the reasoning here suggests Butler-Sloss LJ is referring to the FPCLC of MB throughout, but substituting capacity for competence at various intervals on the understanding that competence is *necessarily* the same as FPCLC, with no obvious rationale for doing so. The consequence is that Butler-Sloss LJ is invoking a psychological and philosophical notion when she is actually applying a legal concept.

Why is this problematic? One may reasonably object that this is not unprecedented and that it may well be desirable for a judge to approach the issue of FPCLC by having regard to a theory of decisional competence. But this is not what Butler-Sloss LJ is seeking to do. She is not presenting an argument in the judgment as to *why* FPCLC should be defined in terms of decisional competence but is instead articulating the legal principle (i.e. FPCLC), which is then applied to the facts of the case. Moreover, to articulate FPCLC in terms of a generic notion of legal capacity (which Butler-Sloss LJ does) is to fail to distinguish the three different forms of legal capacity that exist, only one of which (FPCLC) can correlate to decisional competence.

Children

The common law position in relation to the FPCLC of children under 16 to consent to medical treatment, set out in the landmark case of *Gillick*,[865] has become a quintessential part of English medical law. Although the principle has been subjected to controversial judicial qualification in the intervening years,[866] the basis of the ruling still stands – a child under the age of sixteen has FPCLC to consent to medical treatment "if and when the child achieves a sufficient understanding and intelligence to enable him or her to understand fully what is proposed".[867] The ruling has become known as the test of *Gillick* competence. As Beyleveld and Brownsword observe, the concept of *Gillick* competence is therefore not a measurement of ostensible agency, but an attempt to articulate a standard of decisional competence.[868] Section 43(8) of the Children Act 1989 reflects and extends the *Gillick* principle in statutory form by permitting a child to refuse a medical or psychiatric

[865] *Gillick v West Norfolk and Wisbech Area Health Authority* [1986] A.C. 112.

[866] In particular, see *Re R (A Minor) (Wardship: Consent to Treatment)* [1992] Fam. 11 and *Re W (a minor: medical treatment)* [1992] 4 All ER 627.

[867] *Gillick v West Norfolk and Wisbech Area Health Authority* [1986] A.C. 112, p. 188–189 (Lord Scarman).

[868] Beyleveld and Brownsword (2001: 133, n. 24).

examination as part of a child assessment order provided she is of "sufficient understanding".[869]

Gillick represents a departure from the status-based approach to ascertaining FPCLC in favour of the functional approach. The judgment in *Gillick* delivered by Lord Scarman implicitly reveals a willingness to allow standards of FPCLC in relation to children to be influenced by the same kind of understandings of the decisional abilities of children that we considered in Chapters 5 and 6. It follows that FPCLC to consent to treatment in relation to children is also partly informed by a decision-relative account of decisional competence. However, this rejection of the status test is only partial. Despite espousing of the autonomy of mature children, *Gillick* nonetheless represents a risk-relative standard of decisional competence and an outcome test of FPCLC. This is because it still permits the consent of a parent or the decision of the courts to override any refusal of treatment given by the child. Although the English courts have acknowledged the tension between, on the one hand, the logical symmetry of a decision and, on the other, employing a risk-relative theory of decisional competence with young people,[870] this remains so far unresolved.

In *Re R*, Lord Donaldson considered that the parental right to consent to treatment that *Gillick* circumscribed was to determine the legal approach to a mature minor, but that a parental consent could still be valid in the light of the child's refusal, where it was in the best interests of the child to have that treatment.[871] The decision in *Re R* also precludes fluctuating competence. The court ruled that a determination of FPCLC to consent should be made only after having regard to the general condition of the young person, rather than as a result of an assessment of decisional competence at a point when she was competent.[872] This is because any judgment of decisional competence that may follow would be unrepresentative of her usual level of decisional incompetence.[873] The court's reasoning is compatible with PTDCJ

---

[869] s. 44(7) reiterates the principle, which deals with orders for emergency protection of children and in Schedule 3, which deals with supervision orders. In the former case, the child may refuse to submit to a psychiatric examination or other assessment if she possesses "sufficient understanding". In the latter case, a supervision order is not required if the child has "sufficient understanding" to consent to co-operation. We should note however, that this scope for FPCLC for refusal is circumscribed by the conditions set out in the legislation and does not extend to normal treatment situations, where FPCLC is ascribed to mature children for consent and not refusal.

[870] As acknowledged by Lord Balcombe in *Re W (A Minor) (Medical Treatment: Court's Jurisdiction)* [1993] Fam. 64 at p. 88.

[871] *Re R (A Minor) (Wardship: Consent to Treatment)* [1992] Fam. 11, p. 24.

[872] *Re R (A Minor) (Wardship: Consent to Treatment)* [1992] Fam. 11, p. 24 (Lord Donaldson).

[873] Referring to the facts of the case, Lord Donaldson said: "Even if she [R] was capable on a good day of a sufficient degree of understanding to meet the *Gillick* criteria, her mental disability … was such that on other days she was not only '*Gillick* incompetent', but actually sectionable" (*ibid.*: 26). Assuming that he is correct about this, then under Gewirthian precautionary reasoning, R should be considered a societally incompetent agent, for whom the treatment is necessary in order to protect her from causing unintended harm to herself or others.

insofar as it requires that a test of decisional competence reflects whether the decisional competence in question is a consistent feature of the individual, but is incompatible in the sense that it operates within the context of a risk-relative theory of competence.

The position of FPCLC in respect of 16–17 year-olds to consent to treatment is found in s. 8(1) of the Family Law Reform Act 1969. The section provides that consent to surgical, medical or dental treatment on behalf of the young person should be given effect and supplementary consent need not be sought from the young person's parents.[874] This statutory presumption of FPCLC is indicative of a legal presumption of decisional competence, but one which is subject to the qualification that it is still possible for someone with parental responsibility or the court to override refusal.[875] Since the MCA 2005 came into force, however, the test of capacity set out in s. 3 will apply to young people above the age of sixteen to consent to treatment and research that is covered by the Clinical Trials Regulations. Moreover, as we acknowledged earlier, an adult for the purposes of the Clinical Trials Regulations is defined as someone aged sixteen and above. This may create a tension with the existing case law, as s. 2(3) of the Act states that age cannot be determinative of capacity.[876] It also gives rise to a bizarre current situation whereby a sixteen year-old cannot refuse treatment on her own behalf but can consent to research.

In one sense, s. 8(1) is an extension of risk-relative *Gillick* competence. However, there is an important difference. Whereas we can describe the effect of *Gillick* more accurately as a partial instantiation of a demonstrable risk-relative competence, insofar as the competence must be proved to the satisfaction of the assessor, both the FPCLC of 16–17 year olds to consent to treatment and FPCLC of adults for the same are instances of a presumed decisional competence. In both cases, this is defeasible when the individual in question is found through subsequent examination to lack the cognitive requirements for decisional competence in that regard. The difference between the FPCLC of adults and the FPCLC of 16–17 years olds is that the latter do not also have a rebuttable presumption to refuse. Therefore, the theory of decisional competence that informs the FPCLC of sixteen and seventeen years olds is as asymmetrical as that which informs the FPCLC of young people under sixteen.

---

[874] s. 8(2) states: "The consent of a minor who has attained the age of sixteen years to any surgical, medical or dental treatment which, in the absence of consent, would constitute a trespass to the person, shall be as effective as it would be if he were of full age; and where a minor has by virtue of this section given effective consent to any treatment it shall not be necessary to obtain any consent for it from his parent or guardian."

[875] *Re W (A Minor) (Medical Treatment: Court's Jurisdiction)* [1993] Fam. 64, p. 84 (Lord Donaldson).

[876] Pattinson (2006: 159) also makes this observation.

This view is endorsed in respect of 16–17 year olds in *Re W (a minor) (medical treatment: court's jurisdiction)*,[877] where it was held that the court could exercise extensive powers of wardship to act in the best interests of the young person. Lord Donaldson held unequivocally that it is not possible for the refusal of any person under the age of eighteen to override the consent of someone with parental responsibility or the court.[878] Similarly, Lord Balcombe looked at the wording of s. 8(3) of the Act and concluded that it was not possible to construct an interpretation to provide for an absolute right of refusal.[879] However, Lord Donaldson intriguingly supposed a difference between medical law and medical ethics, which he took to mean that an abortion performed on a sixteen or seventeen year old without her consent but with the consent of the parents would be unlikely to go ahead, unless it was in the young person's best interests.[880] This suggests that positive law is not the only factor which judges are prepared to consider when a tension arises between FPCLC and decisional competence.

Of course, given our earlier recognition that FPCLC need not be construed in terms of decisional competence, it is theoretically possible to frame a test of FPCLC in asymmetrical terms without incurring internal incoherence. There is nothing inherent in the definition of legal capacity that requires FPCLC to attach to all possible options in a given decision making context. The absence of a necessary connection between FPCLC and decisional competence in English law emphasizes this. However, this is a much less compelling explanation for *Gillick*, because of the ubiquitous association of the word 'competence' in academic debate as well as by the judiciary with the *Gillick* standard. If one takes the view argued for in Chapter 1 that risk-relative decisional competence is incoherent, then the ruling in *Gillick* and s. 8(1) of the Family Law Reform Act 1969 are not tests of decisional competence at all but rather choice-relative ascriptions of FPCLC which follow an outcome test approach. Both are tests of FPCLC which draw upon notions of decisional competence but whose conceptual structures ultimately defy the nature of decision-making. Indeed, it would have been more accurate had the *Gillick* test instead been named '*Gillick* capacity'.

Problems Surrounding the Use of Capacity

One might object that the British Government's choice to frame legal capacity in the language of mental capacity through the MCA 2005 is confusing. If one recalls the previous distinction, whereby legal capacity was restricted to legal discourse and mental capacity was restricted to medicine and psychiatry, this objection is

[877] [1993] Fam. 64.

[878] *Ibid.* at p. 84.

[879] *Ibid.* at p. 86.

[880] *Ibid.* at p. 79.

underscored.[881] However, given the appreciable extent to which FPCLC is defined in terms of decisional competence in English law, then using mental capacity as a synonym for FPCLC might be construed as simply describing the same relationship by using different terminology.[882] Understood in these terms, this is merely an innocuous substitution.

This would be a hasty conclusion to reach. Let us first consider the prospects of mental capacity as a new legal concept. Clearly, from the nomenclature of the Act, mental capacity is used in preference to legal capacity. This may have been as a result of a deliberate political choice to emphasise that the Act is seeking to legislate for the position of those who are commonly referred to outside of legal circles as lacking mental capacity. It is clear from the Act that all persons without mental capacity will not have FPCLC under the Act and that some of these persons with mental incapacity, by virtue of having created an LPA before the onset of their incompetence, will have legal capacity exercised on their behalf through their surrogate (the second sense of legal capacity elaborated in Chapter 1).[883]

However, this is not the same as claiming that all persons without FPCLC will also lack mental capacity, on which the Act is silent. Since mental incapacity is defined in s. 2 and 3 of the Act in terms of criteria for decisional incompetence, it is highly foreseeable that there will be groups of individuals who would meet the standard of decisional competence, but who, due to countervailing legal principles in statute and common law, do not have the relevant FPCLC to come within its ambit. Examples would be decisionally competent fourteen or fifteen year olds who wish to consent to medical research on their own behalf or sixteen and seventeen year olds who wish to refuse medical treatment without the risk of that refusal being overruled by their parent/guardian or by the courts.[884] Thus, the presumption in favour of mental capacity set out in s. 1(2) of the MCA 2005 does not cover all individuals who are capable of making decisions and will be circumscribed by other legal principles. Moreover, as the British Medical Association and the Law Society have observed, mental capacity can mean something different to each

---

[881] Law Commission of England and Wales (1991: 19).

[882] This convergence of terminology has had a long history in the proposals for law reform that led up to the passing of the Act (Law Commission, 1993, 1995; Lord Chancellor's Department, 1997, 1999; Department of Constitutional Affairs, 2003).

[883] 'LPA' refers to Lasting Power of Attorney, an expanded class of legally authorised surrogate decision-makers created by the MCA 2005 (replacing the old 'enduring power of attorney'). Section 11(7)(c) of the Act allows an LPA to give or refuse consent to the continuation or carrying out of treatment by a person providing health care for a decisionally incompetent individual. In respect of medical research, Schedule 1 Part 5 of The Medicines for Human Use (Clinical Trials) Regulations 2004 already allows for a legal representative to consent to research participation on behalf of a decisionally incompetent individual where the research is being conducted into a "life-threatening or debilitating condition from which the subject suffers" (Part 5, s. 11).

[884] Levine (1988: 261) identifies this distinction by employing the terms 'de jure' and 'de facto' incompetence, where 'de jure' incompetence is an absence of FPCLC to decide (even though the person may in fact have decisional competence) and 'de facto' incompetence is an absence of decisional competence, which will usually be accompanied by an absence of FPCLC.

profession involved in assessing it.[885] There is, then, the potential for mental capacity and legal capacity to be in practice informed by different standards, even though the Act suggests otherwise.

Towards Supported Decision-Making?

A principal advantage of supported decision-making is that it maximizes the extent to which cognitively vulnerable individuals both possess and exercise legal capacity, as opposed to having legal capacity exercised through a surrogate decision-maker. In the Joint Committee Report on the Draft Mental Incapacity Bill (the precursor to the MCA 2005), the Making Decisions Alliance (a lobbying organisation which represents a variety of support groups) pointed out that the then Draft Mental Incapacity Bill perceived decisional competence as either fully present or fully absent. It would be preferable, they insisted, that steps should be taken to ensure that decision-making is supported through accessible information and communication.[886] This parallels the argument I have made earlier in this book.

There is a requirement in s. 1(3) of the MCA 2005 that an individual should not be treated as incapable of making a decision "unless all practicable steps to enable him to do so have been taken without success". No case law has yet considered the question of what constitutes 'practicable steps' due to the Act only recently having fully come into force. Similarly, the Human Tissue Act 2004 makes provisions relating to "appropriate consent" for the removal and storage of tissue from adults and children who are decisionally competent, although nowhere in this Act is there reference to steps that must be followed to preserve or enhance decisional competence. However, the MCA 2005 Code of Practice does provide some suggestions, such as establishing how the individual usually communicates, using a qualified interpreter, employing 'mechanical devices' (such as a voice synthesizer), engaging in non-verbal communication or offering speech or language therapy as examples of assistance in relation to maximizing decisional competence.[887] The core objectives of supported decision-making could fall within a broad interpretation of any of these examples of assistance.

By way of comparison, s. 1(6) of the equivalent Scottish provision, the Adults with Incapacity (Scotland) Act 2000 stipulates that:

> a person shall not fall within this definition [the definition of incapability of making a decision] by reason only of a lack or deficiency in a faculty of communication if that lack or deficiency can be made good by human or mechanical aid (whether of an interpretative nature or otherwise).

Although s. 1(6) does not suggest that the Scottish legislation is premised necessarily on a concept of supported decision-making, the wording of this section strongly indicates that supportive interventions are of utmost importance when a determination

[885] British Medical Association and the Law Society (2004: 4).

[886] Joint Committee on the Draft Mental Incapacity Bill (2003: 24, para. 73).

[887] Department of Constitutional Affairs (2007, para. 3.11 and para. 2.7).

of mental capacity is at stake. Indeed, the emphasis upon remedying shortcomings of communication by an unspecified range of interpretative devices would seem to catalyse such supportive interventions. If a permissive interpretation of the duties upon physicians and researchers is adopted in relation to the MCA 2005 – and the Scottish legislation may prove a persuasive referent for the English courts – then there is greater potential for a legal accommodation of supported decision-making in English law.

It is important to emphasise that the MCA 2005 could not accommodate supported decision-making by way of the role of the Independent Mental Capacity Advocate (IMCA).[888] The function of the IMCA is qualitatively different in two respects from the concept of supported decision-making. First, the IMCA pertains only to individuals who have already been found to be incompetent to make decisions of that particular type. It does not encompass individuals with questionable capacity who may still be able to make decisions of that type. The use of IMCAs is restricted to scenarios where the person for whom the intervention is designed to benefit does not have a Lasting Power of Attorney or a Deputy appointed. Unlike the supported decision-making approach, then, the Advocate becomes operative only where an individual already has been found to lack decisional competence and where no other authorised surrogate exists.

Second, the role of the Independent Mental Capacity Advocate is limited to certain specified scenarios: provision of serious treatment by an NHS body and provision of accommodation by an NHS body or by a local authority.[889] This adds weight to the terminological preference for 'supported decision-making' in favour of 'advocacy', which has a distinct legal meaning as a result of the implementation of the MCA 2005. The former more easily accommodates the idea of provision of information and advice to an adult who is still competent, in spite of experiencing cognitive vulnerability.[890]

The Significance of European Regulation of Capacity

Recommendation No. R (99) 4 (1999) of the Committee of Ministers to Member States on Principles Concerning the Legal Protection of Incapable Adults[891] emphasises the preservation of decisional competence as the central value that should underpin

[888] Set out in s. 35–41 of the MCA 2005.

[889] s. 37–39.

[890] In the *Consultation on the Independent Mental Capacity Advocate Service* (UK Department of Health, 2005), the Government signalled their willingness to consider expanding the service to include individuals who are not suffering from mental incapacity for the purposes of the Act and intimated that the service could expand to include research (*ibid.*: 38). However, the significant costs associated with expanding this service and the lack of enthusiasm for it in the consultation means that this was not carried forward in the Department of Health's 2006 response to the consultation or in The Mental Capacity Act 2005 (Independent Mental Capacity Advocates) (Expansion of Role) Regulations 2006. It is also worth noting that research does not fall within the remit of independent mental health advocates recently created by s. 30 of the Mental Health Act 2007 (but not yet in force at the time of writing).

[891] Council of Europe Committee of Ministers (1999).

any approach to dealing with individuals with mental impairment. It was devised by a group of specialists on Incapable Adults (CJ-S-MI) set up by the Council of Europe in 1995.[892] As a Recommendation, it has no binding force within EU law, but it embodies nonetheless a clear articulation of decisional competence and its relationship with FPCLC:

> *"Principle 3 – Maximum preservation of capacity:*
>
> 1. The legislative framework should, so far as possible, recognise that different degrees of incapacity may exist and that incapacity may vary from time to time. Accordingly, a measure of protection should not necessarily result automatically in a complete removal of legal capacity. However, a restriction of legal capacity should be possible where it is shown to be necessary for the protection of the person concerned.
> 2. In particular, a measure of protection should not automatically deprive the person concerned of the right to ... consent or refuse consent to any intervention in the health field, or to make other decisions of a personal character at a time when his or her capacity permits him or her to do so."[893]

These principles are close in nature and application to the MCA 2005 and to PTDCJ. Paragraph 1 defines 'incapacity' as a task specific or decision-relative quality rather than as a lack of agency competence. This paragraph also attempts to distinguish mental capacity from legal capacity. This is welcome as it seeks to avoid the potential for conflation that stems from the unqualified use of 'capacity'. Since there appears to be a strong connection between the decision-relative use of 'capacity' and 'legal capacity' (understood as FPCLC), a lack of FPCLC can be taken to amount to a legal recognition of the absence of a task specific or decision-relative competence rather than as an inability to exercise any FPCLC at all.

From the perspective of PTDCJ, this is a justifiable interpretation of FPCLC, as it is only withheld in spheres of activity where decisional competence is not apparent. Furthermore, the measures of protection that are proposed appear to be proportionate to the task and decisional inabilities that a person may possess. This is very similar to how we must treat societally incompetent agents under the PGC. Although there may be grounds for additional protections (the words "not *automatically*" in paragraph 2 suggests there is some scope for depriving an individual of FPCLC if on the evidence, the individual lacks decisional competence in this respect), the overriding consideration appears to be respect for individual autonomy.

In paragraph 2, there is an explicit recognition that the protection owed to an individual with an unrelated specific task or decisional incompetence need not render the individual incapable of being able to make decisions about medical interventions, providing that she does possess decisional competence in this respect. We do not need to speculate as to the meaning of "interventions in the health field", as earlier in the Recommendation, the definition is framed to include both treatment

[892] Jansen (2000: 333).

[893] Part 2, Principle, 3, s. 1–2.

and research.[894] The inclusion of the phrase "consent or refuse" substantiates our earlier claim that the use of 'capacity' in the Recommendation is grounded upon a decision-relative concept of competence. It also provides support for the view that we should as far as possible respect the competences of individuals at the time they are exhibited, even if at previous or subsequent times they are absent.

Later in the Recommendation, Principle 22 states that provided an adult is capable of giving informed consent, even where an adult is the recipient of protective measures, the intervention may only be carried out with his or her consent.[895] But how can we know, under the Recommendation, whether or not an individual is indeed decisionally competent to give or withhold consent? Principle 12 stipulates, "[T]here should be adequate procedures for the investigation and assessment of the adult's personal faculties,"[896] although no reference is made to the form or substance that the competence assessment would take or the standards to which it should have recourse. This devolves responsibility for devising or choosing a competence assessment measure to an unnamed authority and appears not to require consistency in approach. Without any elaboration of the test which may be used to assess competence, the Recommendation's aim to preserve capacity as long as possible is compromised.

Other instruments offer much less scope for furthering our understanding of the regulatory meaning of 'capacity'. The two principal instruments that seek to regulate research on human participants in the EU, the European Convention on Human Rights and Biomedicine (ECHRB),[897] along with its Additional Protocol on Biomedical Research[898] and the Clinical Trials Directive (Directive 2001/20/EC),[899] emphasise special protections owed toward the participation of adults with mental disorder and children. However, none discuss the issue of the assessment or preservation of decisional competence or FPCLC. Indeed, the phrasing of the ECHRB and the Clinical Trials Directive suggests that both approach the issue of the participation of adults with mental disorder *not* from a standpoint that requires decisional competence be assessed before the issue of surrogate decision-making arises, but from one that assumes surrogate decision-making as the appropriate starting point in relation to their participation. Where brief reference is made to cases of doubtful competence in Article 14(3) of the Additional Protocol to the ECHRB on Biomedical Research, there is simply a statement that "arrangements shall be in place to verify whether or not the person has such capacity". There is no attempt to expand upon what form these arrangements might take or to what criteria these arrangements must adhere.

---

[894] Principles – Part 1 – Scope of application, para. 5.

[895] Part 5, Principle 22, s. 1.

[896] Part 3, Principle 12, s. 1.

[897] Council of Europe (1997). As of 1 March 2008, the UK still has not signed up to the European Convention on Human Rights and Biomedicine.

[898] Council of Europe (2005).

[899] Directive 2001/20/EC, OJ L 121, 1.5.2001.

On other occasions, we can witness internal incoherence within the same guidance on capacity. In the 2000 Council of Europe White Paper on the protection of the human rights and dignity of people suffering from mental disorder, especially those placed as involuntary patients in a psychiatric establishment,[900] it was suggested that the concept of mental capacity in Europe was inadequately articulated and needed development. Interestingly, the Council of Europe thought such mentally disordered persons should be protected from involvement in non-therapeutic clinical trials even if they consent to them.[901] It would appear that the White Paper saw scope for development of FPCLC more in terms of constraint, in which paternalistic considerations could trump the moral authority of a decisionally competent consent. This is redolent of the status test of FPCLC outlined at the beginning of the chapter, which if used, would circumscribe the autonomy of decisionally competent patients, reconceptualizing the best interests test as one which applies to competent cognitively vulnerable research participants as well as incompetent individuals. This would be incompatible with the approach taken under PTDCJ, because unless the research had not been subjected to ethical review, had failed such review or was ethically deficient in some other way (according to standards consistent with the PGC), decisional competence to consent should otherwise be a sufficient condition for participation.[902]

The emphasis that the Convention and the Directive place upon research relating to a condition experienced by the individual concerned is clearly motivated by an intention not to exploit their vulnerability. On the face of it, this is laudable. Nonetheless, there is a pervasive shortcoming which follows from the failure to distinguish research participants into individuals capable and incapable of giving consent without first addressing the issue of competence assessment. Equally, there is no reference to why preserving capacity should be valuable and how it should be undertaken in the context of medical research. Due to a silence on the issue of assessing and judging competence found in all these instruments surveyed, along with any meaningful reference to supported decision-making, the status of decisional competence to consent and its ability to inform ascriptions of FPCLC is frustrated.

Ethical Guidance on Decisional Competence in Biomedical Research in the UK

The ethical guidelines that have emerged on the protection of human participants in clinical research in the UK since the beginning of the 1990s have preceded legal discussion of these issues. Typically, there are three forms of ethical guidance in

[900] Council of Europe (2000).

[901] p. 34 paragraph 32. This reiterates Article 5, paragraph 3 of Recommendation No. R 83(2) (1983) concerning the legal protection of persons suffering from mental disorder placed as involuntary patients, without any discernible justification.

[902] That is, assuming the existence of no public wrong. See Beyleveld and Brownsword (2007a: 356–357).

research, that which covers good practice in research generally, that which pertains to adults and that which pertains to children.

One of the first guidelines issued during this period was *The Ethical Conduct of Research on the Mentally Incapacitated*, published by the Medical Research Council (MRC) in 1991.[903] The MRC revised this guidance as *Medical research involving adults who cannot consent* in 2007, to reflect the legal developments brought about by the MCA 2005 and the Clinical Trials Regulations 2004.[904] In the 1991 version, the MRC concluded in paragraph 8.1 that, "Many people with mental impairment or disorder are able to consent to their inclusion in research provided care is taken to explain it to them".[905] This represented a progressive attitude towards the participation of adults with mentally disorder or intellectual disability in medical research, pre-empting the ethos of the 1999 Council of Europe Recommendation. The same statement does not appear in the 2007 guidance, although echoing the endorsement of capacity preservation in the MCA 2005, the revised guidance states: "Capacity is present if the person only has a difficulty with communication that can be overcome with human or mechanical assistance".[906] The guidance goes on to address approaches to supported decision-making in the context of vignettes, recognising the importance of eliciting the decisional competence of potential participants with mental disorder and intellectual disabilities as far as possible.[907]

In 2001, the UK Department of Health published its guidance on informed consent to medical examinations or treatment.[908] This also includes discussion of consent to research. It stipulates:

> 2.6. Care should also be taken not to underestimate the capacity of a patient with a learning disability to understand. Many people with learning disabilities have the capacity to consent if time is spent explaining to the individual the issues in simple language, using visual aids and signing if necessary.[909]
>
> [....]
>
> 15. ... when seeking consent from patients for research purposes ... "particular care" should be taken to ensure that possible research subjects have the fullest possible information about the proposed study and sufficient time to absorb it. Patients should never feel pressurised to take part, and advice must be given that they can withdraw from the research project at any time, without their care being affected. If patients are being offered the opportunity to participate in a clinical trial, they should have clear information on the nature of the trial.[910]

This guidance reflects a belief in the maximization of FPCLC to consent through the role of education and information provision. The specific reference to the assessment

---

[903] Medical Research Council (1991).

[904] Medical Research Council (2007).

[905] Medical Research Council (1991: 22).

[906] Medical Research Council (2007: 9).

[907] Medical Research Council (2007: 12–13).

[908] UK Department of Health (2001a).

[909] UK Department of Health (2001a: 5).

[910] UK Department of Health (2001a: 9).

of decision-relative decisional competence supports PTDCJ and the tenor of the 1999 Recommendation, and shares affinities with the MCA 2005. Furthermore, the emphasis placed upon learning disabilities or factors that may lead to temporary decisional incompetence as constituting no necessary impediment to FPCLC reinforces the possible coexistence between the two. There is also a clear recognition that these principles should be applied with even greater care to research as to treatment. This reflects the common lack of personal benefit in the case of research, and the greater level of comprehension required in order to consent.

Specifically in relation to research, the General Medical Council in its 2001 draft guidance entitled *Medical Research: The Role and Responsibilities of Doctors*[911] explicitly states, "most adults with mental illness or disorders are competent to decide whether or not to participate in research".[912] The final version of the guidance, published in 2002, goes on to refer to assessing decisional competence to consent with vulnerable adult research participants. According to paragraph 43, vulnerable adults include those experiencing mental disorder or intellectual disabilities, along with institutionalized adults and frail elderly persons. The GMC acknowledge that consent sought under the pressures from the health care professionals or institutions with which they have contact might compromise the validity of that consent. However, these adults can be "competent but vulnerable" and that to omit research with vulnerable groups simply because of the ethical issues to which it gives rise may engender discrimination.[913] They go on, in paragraphs 44 and 46 to recommend approaches to dealing with members of these groups sensitively:

> 44. Careful consideration should therefore be given to involving vulnerable adults in research, and particular attention should be given to the consent process, ensuring that they have sufficient information provided in a suitable format, and enough time to consider the issues. You should give consideration to their vulnerability and difficulties they may have in understanding or retaining information. You may need to encourage them to seek the help of a relative/close friend, support worker/advocate. You should proceed with the research only if you believe that the participant's consent is voluntary and based on an understanding of the information they have been given.[914]
>
> [....]
>
> 46. Where participants have difficulty retaining information, or are only intermittently competent to make a decision, you should provide any assistance they might need to reach an informed decision. You should record any decision made while they were competent, including the key elements of the consultation. You should review any decision made whilst they were competent at appropriate intervals before the research starts, and at intervals during the study, to establish that their views are consistently held and can be relied on.[915]

---

[911] General Medical Council (2001).

[913] General Medical Council (2001, para. 30). Curiously, this statement was not adopted in the final version of the guidance (General Medical Council, 2002), but there is nothing in the final version that would contradict this statement.

[913] General Medical Council (2002, para. 43).

[914] General Medical Council (2002, para. 44).

[915] General Medical Council (2002, para. 46).

The GMC guidance indicates that grounds exist for treating cognitively vulnerable adults as decision-makers in their own right, by virtue of consistency in their decision-making capabilities and the effects of assistance on their understanding, which tacitly endorses the methods we examined in Chapter 6. However, this approach towards adults is at odds with the legal position of children under sixteen who are "intermittently competent to make a decision", since, as we saw earlier, they are not considered to possess decisional competence for a sufficient length of time to be considered *Gillick* competent.[916]

This anomaly raises the issue of what could justify the participation of adults with fluctuating decisional competence to consent but not children. A plausible response is the need for a consistent set of values to underpin decision-making, emphasised in the competence theories of Wicclair, Grisso and Appelbaum considered in Chapter 1. Whereas adults with intermittent decisional competence are more probably in possession of these values, an older child or adolescent with fluctuating capacity is less probably in possession of them. This is broadly reflective of the evidence for task and decisional competences that typically differentiates adults and older children, with the adult more probably having a wider range of task and decisional competences and to a greater degree, and the older child more probably having fewer task and decisional competences, and to a lesser degree.

The Medical Research Council issued guidance in 2004, entitled *Medical Research Involving Children*, which replaced its earlier guidance on the same issue.[917] It recommends:

> Seeking consent is not a single response but a *process*. The child should be provided with information appropriate to his or her increasing ability to make decisions about complex and serious issues. It is helpful for researchers to produce child-friendly information in a form appropriate for the relevant age groups – this could make use of pictures or videos. More than one version may need to be produced if research covers a wide age range, such as eight-18 years.[918]

These recommendations present potentially effective solutions with which to elicit a young person's understanding. Unlike the previous guidance we have examined, they suggest examples as well as general strategies for achieving this. The MRC's recommendation is predicated on a belief that young people can and should become involved in making decisions about themselves, and thus meshes with a number of approaches that seek to encourage young people to develop their own autonomy, without simply being abandoned to it. Unlike the previous edition of the guidance, there is also an explicit recognition that the decisional abilities of children change vastly from early childhood to adolescence and that forms of supported decision-making need to be tailored to the needs of the age group. One hopes that the empowering and supportive model of developing decisional competence proposed here will find widespread acceptance.

---

916 *Re R (A Minor) (Wardship: Consent to Treatment)* [1992] Fam. 11.

917 Medical Research Council (2004).

918 Medical Research Council (2004: 34).

Summary

Decisional competence has no legal status in English law. Although British judges and legal academics frequently invoke the term 'competence' in discussions of cases involving 'capacity to consent', it would appear to be no more than as a synonym for a poorly defined notion of legal capacity. It is poorly defined because there is no attempt to distinguish between the three types of legal capacity which we introduced in Chapter 1. This has given rise to a conflation of competence and capacity in English law, as we cannot ultimately know in what sense competence is being invoked by the unqualified term 'capacity'. This risks jeopardizing legal clarity and consistency.

The emergence of 'mental capacity' as a new legal concept does suggest a close link between a notion of FPCLC defined in terms of mental capacity (which the MCA 2005 embodies) and decisional competence. However, there is no necessary connection between the two. There remains a possibility for FPCLC and mental capacity to be assessed by different standards in spite of the correspondence between s. 2(a–d) of the MCA 2005 and decision-relative decisional competence. This, along with the silence surrounding whether those persons without FPCLC should therefore be considered as not having mental capacity suggests that mental capacity cannot be read entirely as a synonym for decisional competence.

Whilst the definition of mental capacity in the Act maintains internal coherence, when applied to these broader questions raised by factors outside of the Act, questions arise about its external coherence. If one is seeking clarity and consistency in the law, it would have been wiser to have renamed mental capacity in the Act as FPCLC, which would minimize ambiguity and also admit more clearly and consistently of the distinction between FPCLC, delegable legal capacity and fiduciary legal capacity. Insofar as individuals could still have decisional competence to consent when they lack FPCLC to do so, FPCLC should be formulated so that there is a necessary conceptual relationship between the two. A second legal category, delegable legal capacity, should be created to accommodate surrogate decision-making, and the third, fiduciary legal capacity, to refer to decision making by parents or legal guardians on behalf of young children.

Legislating for decisional competence to consent to medical research is a first step towards clarity and consistency in the law. But this needs to be accompanied by a higher profile for supported decision-making than is currently accorded in the MCA 2005 to ensure that it is seen as a mandatory part of the consent seeking process in medical research, rather than a more nebulous form of 'good practice'. In this chapter, we have witnessed the beginning of moves in this direction, within the MCA 2005 Code of Practice, the 1999 Council of Europe Recommendation and in ethical guidance from the MRC and the GMC. Ultimately, however, success in meeting these aims depends also upon maintaining and cultivating attitudes of compassion, ethical integrity, fairness and diligence amongst those responsible for making judgments of competence and supporting decision-making in both the medical and legal fields. Legislating for this is much more problematic.

Chapter 8
Cognitive Vulnerability and Consent to Biomedical Research in the United States

That much bioethical literature on competence has emanated from the US makes the regulatory position worth studying. In this final chapter, we will consider to what extent there are, if at all, coherent standards of decisional competence to consent and first person contemporaneous legal competence (FPCLC) in the United States both at a federal and state level, and if so, whether they are consistent with precautionary task or decisional competence judgment (PTDCJ). The chapter begins by establishing the meaning of competence and capacity in the United States in order to investigate whether FPCLC is synonymous with decisional competence, and the extent to which cognitive vulnerability militates against this. We will go on to consider how federal and state laws approach competence and identify selected policy proposals that have addressed the competence of cognitively vulnerable participants in medical research. Where relevant, we will consider the prospects for supported decision-making under existing laws and ethical guidelines.

The Meaning of Competence and Capacity in the United States

In the United States, competence is primarily a legal concept. Capacity, instead, refers to a clinical judgment of a present or future level of decision-making ability that may have no legal consequences.[919] Competence is not always completely specified and criteria may vary between jurisdictions.[920] Its practical meaning is therefore harder to establish.

It may come as no surprise, then, that the US regulatory model is limited. Meslin argues, "[t]he prevailing model of informed consent in research involving human subjects ... reflected in US guidelines and regulations, is best understood as one in which a competent, otherwise healthy adult is invited to participate in a clinical trial of new medication".[921] Berg et al. go further, arguing:

919 Berg et al. (1996: 348–349, n. 14).

920 Berg et al. (1996: 347–349).

921 Meslin (2001: 151).

> presently there is a lack of both an authoritative framework for thinking about legal competence and clear standards for determining it. Cases and statutes generally lack sufficient analysis of competence and its different elements ... Even where a statute articulates a standard of competence, it is often vague and provides little guidance for those who must apply it.[922]

The absence of comprehensive standards of decisional competence suggests that the legislature and judiciary assume that anyone who consents to the satisfaction of the assessor thereby will be competent in this respect. However, without first establishing decisional competence, particularly where there are grounds for calling such competence into question, the moral and legal validity of any resulting consent is undermined. To provide relevant information to the potential participant in an environment free from coercion or undue influence cannot compensate for undeveloped decisional competence.

The legal marginalization of decisional competence in the United States is in part explained by the fact that competence assessments made by psychiatrists or other health care professionals are not legally binding. According to Berg and Appelbaum, "in most jurisdictions, only a court can decide whether a person is incompetent".[923] This is extraordinary, given the role of the psychiatrist, physician or researcher is to usually make the primary determination of competence or incompetence, which in most cases will stand unless the individual about whom the judgment is made appeals against this through the courts. Of course, the courts do have an important role in assessing the procedural integrity of competence assessments, particularly if there is dispute about the legitimacy of a judgment of decisional competence. Nevertheless, health care professionals are usually better informed to measure competence to consent than judges, given their specialist training and direct contact with the potential participant. While US law may claim competence as its own concern, so far it appears unwilling to give the competence determinations of healthcare professionals legal standing.

Federal Law

In 1981, federal regulations were introduced in the United States establishing a national legal framework in respect of research on adult human subjects. These regulations comprise the Federal Policy for the Protection of Human Subjects (FPPHS),[924] which were adopted subsequently by sixteen US federal agencies in 1991, and as such have come to be known as the 'Common Rule'.

---

[922] Berg et al. (1996: 348–349).

[923] Berg and Appelbaum (1999: 82, n. 1).

[924] Basic DHHS (Department of Health and Human Services) Policy for the Protection of Human Research Subjects, 45 Code of Federal Regulations Part 46: Revised June 23 2005; Effective June 23 2005.

Subpart A of the FPPHS is the primary source of research regulation in the United States. It applies to all research involving human participants "conducted, supported or otherwise subject to regulation by any Federal Department or Agency",[925] both inside and outside the United States and includes research conducted by Federal civilian employees or military personnel. Within this subpart are the provisions relevant to consent and capacity:

> "§46.116 General requirements for informed consent
> Except as provided elsewhere in this policy, no investigator may involve a human being as a subject in research covered by this policy unless the investigator has obtained the legally effective informed consent of the subject or the subject's legally authorized representative. An investigator shall seek such consent only under circumstances that provide the prospective subject or the representative sufficient opportunity to consider whether or not to participate and that minimize the possibility of coercion or undue influence. The information that is given to the subject or the representative shall be in language understandable to the subject or the representative.
> ... (d) An IRB may approve a consent procedure which does not include, or which alters, some or all of the elements of informed consent set forth in this section, or waive the requirements to obtain informed consent provided the IRB finds and documents that:
>
> (1) the research involves no more than minimal risk to the subjects;
> (2) the waiver or alteration will not adversely affect the rights and welfare of the subjects;
> (3) the research could not practicably be carried out without the waiver or alteration; and
> (4) whenever appropriate, the subjects will be provided with additional pertinent information after participation."[926]

The FPPHS clarifies the requirements for informed consent in meticulous detail. Beyond this, however, there are two main shortcomings. First, whilst there appears to be a presumption that any potential research participant will be decisionally competent, nowhere does it specify the requirements for competence or the grounds upon which it is be assessed or judged.[927] Second, §46.116 part (a) 1–8 and (b) 1–6 are framed in terms of what the researcher must do to discharge her legal duty to the potential participant, without considering that the potential participant herself will usually be disempowered due to lack of medical training, knowledge of the research process, and the common feeling of being overwhelmed by having to absorb a great deal of new information. Although there are numerous legal protections in this section, including the prohibition of terms of consent that waive the participant's legal rights, these requirements do not appear to address the relational inequalities that exist between researcher and participant. Moreover, there is no recognition that a process of obtaining consent must prioritize support and counselling as highly as a specification of how much and what kind of information to disclose.

---

[925] §46.101, a.

[926] §46.116.

[927] Berg and Appelbaum (1999: 82).

It is unclear as to why §46.116 negates the assessment of competence, given its centrality to informed consent. Given that there are few cases in the US that address competence to consent to treatment, and none in respect of research,[928] this is a disappointing omission. Additionally, the wording of §46.116 (d) that permits the IRB to approve a consent procedure that alters or even waives entirely the requirement to obtain informed consent within the context of "public benefit or service programs" generates concern. The kinds of programmes this section is referring to are not disclosed, which increases the amount of interpretative discretion this confers upon the state or local government officials.[929]

The ultimate impression gained from a critical analysis of the FPPHS is one of an intricate framework intended to respond to the concerns of IRBs and investigators by systematising what must be done to absolve them from legal liability in the event of an action in negligence. In other words, the securing of informed consent is a legal formality that does not recognise the inherent inequality in the relationship between the researcher and potential participant and the consequent heightened vulnerability experienced by the latter. The fact that there is no reference to assessing or judging competence, or supporting decision-making, compounds this further.

Federal legislation governing research with children was introduced in 1983. This has since been subsumed as Subpart D of the FPPHS.[930] The definition of a child is a person who has not yet reached, "the legal age for consent to treatments or procedures involved in the research" in the jurisdiction where the research will be carried out.[931] As with Subpart A, the provisions create basic legal duties:

> "§46.402 Definitions.
> … (b) Assent means a child's affirmative agreement to participate in research. Mere failure to object should not, absent affirmative agreement, be construed as assent.
> [….]
> §46.408 Requirements for permission by parents or guardians and for assent by children

(a) In addition to the determinations required under other applicable sections of this subpart, the IRB shall determine that adequate provisions are made for soliciting the assent of the children, when in the judgment of the IRB the children are capable of providing assent. *In determining whether children are capable of assenting, the IRB shall take into account the ages, maturity, and psychological state of the children involved* [emphasis added]. This judgment may be made for all children to be involved in research under a particular protocol, or for each child, as the IRB deems appropriate. *If the IRB determines that the capability of some or all of the*

---

[928] Berg et al. (2001: 266).

[929] It is perhaps worth nothing that since October 2000, the National Institutes of Health (NIH) requires education on the protection of human research participants for all investigators submitting NIH applications for grants or proposals for contracts or receiving awards for research involving human subjects. However, this may be insufficient to educate investigators on the issue of competence, unless consideration is also given to its ethical significance.

[930] Additional DHHS Protections for Children Involved as Subjects In Research, Subpart D, 45 CFR 46 (1983).

[931] *Ibid.* §46.402.

> *children is so limited that they cannot reasonably be consulted or that the intervention or procedure involved in the research holds out a prospect of direct benefit that is important to the health or well-being of the children and is available only in the context of the research, the assent of the children is not a necessary condition for proceeding with the research* [emphasis added]. Even where the IRB determines that the subjects are capable of assenting, the IRB may still waive the assent requirement under circumstances in which consent may be waived in accord with §46.116 of Subpart A.
>
> (b) In addition to the determinations required under other applicable sections of this subpart, the IRB shall determine, in accordance with and to the extent that consent is required by §46.116 of Subpart A, that adequate provisions are made for soliciting the permission of each child's parents or guardian."[932]

The definition and practice of assent is fine for all ethical and practical purposes when dealing with young children, but under the terms of Subpart D, is meant to apply to all individuals who have yet to attain legal majority. As Glantz observes, an individual state may enact legislation (or have authority within its case law) which allows a fifteen year old to consent to a novel therapy as part of a research project, in which case the child in question is no longer a child for the purposes of the FPPHS.[933] However, in order to minimize inconsistency of approach, it is somewhat surprising that there appears to be no provision for case-by-case assessment of decisional competence here, or for a statutory sub-class of young people who are presumed to have FPCLC to consent to research. This reveals an inflexible shortcoming of the legal provision that amounts to a status model of competence by which US law technically places adolescents in the same legal category as young children. Parental legal responsibility is thus the point of departure in decisions about research participation, giving rise to an extensive application of fiduciary legal capacity.[934]

§46.407, which pertains to non-therapeutic research, is similarly problematic insofar as the assent of children and the permission of parents are required without any qualification on the appropriate limits of assent and permission (for instance until such a time as the child achieves a sufficient understanding and maturity to understand what the research procedure involves). As in Subpart A §46.117, Subpart D §46.408 also contains a provision for waiving the assent requirement altogether. The foreseeable circumstances in which this could be justified are broader than in research with adults, particularly with infants and younger children, but there is little discussion of any correlative protection that could be implemented in order to ensure that their comparatively greater vulnerability is not exploited.

---

932 *Ibid.* §46.402, §46.408.

933 Glantz (1994: 123).

934 See Glantz (1994 and 1998) for a fuller discussion of these issues and the circumstances where exceptions may be made. I would like to thank one of the anonymous reviewers who drew my attention to this point.

In addition, there is an attempt to sketch a basis for an appropriate mechanism to substitute parental permission to research for neglected or abused children, but the relevant options available to the IRB are not delineated.[935] One is left wondering whether these options specifically concern forms of surrogate decision-making or extend to counselling and advice on the part of social service professionals. Overall, there is little to commend the structure and content of Subpart D of the FPPHS, apart from the rather empty recognition that the existence of a statutory framework, although possessive of shortcomings, is better than none.

State Law

There is some evidence to support the claim that the prevailing legal conception of competence, at state level at least, is decision-relative rather than risk-relative. In a seminal article, Roth et al. observed of the legal position in the US that a person "could be considered competent for some legal purposes and incompetent for others at the same time. An individual is not judged incompetent merely because he or she is mentally ill".[936] This assertion is endorsed in *State of Tennessee Department of Human Services v Mary C. Northern*,[937] in which it was held:

> Capacity means mental ability to make a rational decision, which includes the ability to perceive, appreciate all relevant facts and to reach a rational judgment upon such facts. Capacity is not necessarily synonymous with sanity … A person may have 'capacity' as to some matters and lack 'capacity' as to others.[938]

It is interesting that in this case 'capacity' is used ambiguously. The court appears to have either substituted capacity as a synonym for FPCLC or is expressing support for assessment of decisional competence by a health care professional on a decision-relative basis. This is confusing, since there is no way of determining that the court is making reference to a standard of legal competence or a non-legal standard of decisional competence. However, the reliance shown by the court upon the New Jersey case of *In Re Quackenbush*[939] suggests that the standard discussed is more probably legal as well as medical. This latter case involved a seventy-two year-old man with intermittent mental lucidity who was considered decisionally competent to refuse treatment of his own gangrenous legs due to his ability to appreciate the health risks involved in doing so. Although the court in *Northern* found against the respondent, reiteration of the emphasis placed in *Re Quackenbush* upon decisional competences enduring even after the onset of mental impairment[940]

[935] §46.408(c).

[936] Roth et al. (1977: 280).

[937] (1978) 563 S.W. 2d 197 (Tenn Ct App).

[938] Todd J, at p. 209.

[939] (1978) 156 NJ Super 282, 383 A 2d 785 (Morris County Ct 1978).

[940] *Ibid.* at p. 788 (Muir, A. J. S. C).

suggests that legal competence in the US, at least at state level, tends towards decision-relativity.

The philosophical origins of competence and capacity in the US reaches back to *Schloendorff v Society of New York Hospital*,[941] in which Cardozo J famously asserted, "Every human being of adult years and sound mind has a right to determine what shall be done with his own body".[942] This principle has now become legendary in symbolising the move away from medical paternalism and towards patient autonomy during the twentieth century. In the context of the time in which it was delivered, this principle was bold and innovative, rejecting the previously unquestioned authority of the medical profession. However, taken literally, this principle denies FPCLC to children, adults with mental disorder and adults with intellectual disability.

Subsequent case law throughout several US states has neither affirmed nor denied this application of the *Schloendorff* principle, but has refined the concept of competence in such a way as to make it unlikely that legal minority or mental impairment would alone be sufficient for a determination of legal incompetence. In *Canterbury v Spence*[943] the court, after reaffirming the *Schloendorff* principle, held that consent is the "informed exercise of a choice"[944] which requires "an opportunity to evaluate knowledgably the options available and the risks attendant upon each",[945] and that "all risks potentially affecting the decision must be unmasked",[946] according to what a "prudent person in the patient's position" would wish disclosed.[947]

If the standard of the prudent patient is to be interpreted as the epitome of the decisionally competent patient, then it would seem to follow that this standard comprises both evaluation and information comprehension as criteria of decisional competence. However, there are problems with this interpretation. The nomenclature of a prudent patient restricts the application of the principle in research settings outside of the therapeutic relationship (for example, where the individual in question is a volunteer). It is also probable that, in the absence of a common law definition of competence, the standard of a 'prudent patient' is liable to highly subjective interpretation by both judges and healthcare professionals when assessing competence.

The cases of *In Re Maida Yetter*[948] and *Kaimowitz v Department of Mental Health of the State of Michigan*[949] both decided in the following year shed light on

---

941 (1914) 211 NY 125.

942 (1914) 211 NY 125, at p. 129.

943 464 F 2d 772 (DC, 1972).

944 *Ibid.* at p. 780 (Robinson J).

945 *Ibid.* at p. 780 (Robinson J).

946 *Ibid.* at p. 787 (Robinson J).

947 *Ibid.* at p. 791 (Robinson J).

948 (1973) 62 Pa. D. & C.2d 619 (Pa.Com.Pl.).

949 (1973) 42 USLW 2063 (Mich Cir Ct).

two divergent legal understandings of competence in the US. *Maida Yetter* concerned a lucid schizophrenic committed to a psychiatric institution who refused potentially life-saving treatment. Considering the context in which the refusal was issued, Williams J opined that, "it is clear that mere commitment to a state hospital for treatment of mental illness does not destroy a patient's competency … mental capacity must be examined on a case by case basis".[950] This view is welcome from the perspective of PTDCJ for two reasons. First, it emphasises that the ongoing experience of mental illness and the environmental factors involved in treatment are not a sufficient basis on which to found a judgment of legal incompetence.[951] Second, through stressing the importance of case-by-case assessment, Williams J appears to be advocating a functional approach towards the assessment of decisional competence, if we understand the distinction between competence and capacity in the US-specific terms elaborated by Berg et al.[952]

The Michigan case of *Kaimowitz*, alternatively, rejected a belief in the enduring possibility of decisional competence to consent to novel psychosurgical interventions during psychiatric detention.[953] Commenting on the requirement of the Nuremberg Code, the court held:

> It is impossible for an involuntarily detained mental patient to be free of ulterior forms of restraint or coercion when his very release from the institution may depend upon his co-operating with the institutional authorities and giving consent to experimental surgery.[954]

Both these rulings rest upon the situational context of competence, in particular in relation to the institutional environment of a psychiatric hospital, and whether it is has a benign or coercive impact upon decisional competence to consent. Whereas in *Maida Yetter*, the court did not consider the institutional environment to erode decisional competence, in *Kaimowitz*, the relational inequality in the relationship between doctor and patient fused with the nature of the procedure in question was sufficient to vitiate the provision of a competent consent.

We can explain the reasoning behind these judgments by reference to a decision-relative account of decisional competence, given the differing levels of competence that the procedures in the two cases require. In *Maida Yetter*, the issue concerned competence to consent to surgical biopsy recommended due to suspected breast cancer, which constitutes treatment. However, in *Kaimowitz*, the issue concerned competence to consent to experimental psychosurgery, which, in spite of putative therapeutic benefits, constitutes research. Because the standard of competence required in cases

---

[950] (1973) 62 Pa. D. & C.2d 619 (Pa.Com.Pl.) at p. 623 (Williams J.).

[951] Nevertheless, the liberty-limiting dimensions of the institution may impose restrictions upon the ways in which task and decisional competences could be expressed. Certain task and decisional competences may deteriorate or lie dormant because of institutionalisation. This could feasibly include the decisional competence to consent to research.

[952] Berg et al. (1996: 348, n. 14).

[953] Although the reasoning of the Court appeared to vacillate at certain points. See Shapiro (1974: 316–317, n. 275).

[954] (1973) 42 USLW 2063 (Mich Cir Ct) at p. 27, quoted in Kennedy and Grubb (2000: 763).

of research is often greater than in cases of treatment, it is a defensible presumption that an institutionalised patient is competent to consent to *fewer* types of medical research than she is to types of treatment. This is due to both the higher level of the five general constitutive abilities of decisional competence required in research and the potential for the consent process in research in institutional settings being subject to covert forms of manipulation or deception.[955]

Kennedy and Grubb criticize the decision in *Kaimowitz* as over-protective.[956] This has some justification. It should not be due to a presumption that an institutionalised patient is, by definition, decisionally incompetent or lacks FPCLC to offer consent to research.[957] Rather, the presence of decisional competence or incompetence to consent to research should first be evaluated in terms of the personal characteristics of the patient (i.e. whether she is confident and assertive or withdrawn and submissive within the institutional setting) and the extent to which the institution allows and encourages patients to make their own choices more generally.

Certain states have developed their own staged tests of decisional competence as a means to inform ascriptions of FPCLC. In the New Jersey case of *In Re Schiller*,[958] the court ruled that the standard of competence brought to bear upon patients in the context of medical treatment should be that of understanding the condition, nature and effects of the procedure, and the risks associated with its performance and non-performance.[959] This four-stage test has parallels with the *Gillick* standard of capacity to consent, except here the test is elaborated in relation to adults. In order for this test to be applied to research, an understanding of the condition would need to be substituted with an understanding of the aims and objectives of the research procedure. As we saw in Chapter 1, the inclusion of risk appreciation as a criterion of competence need not lead to an asymmetrical application of competence, provided that appreciation of risk is accounted for at the level of ascertaining decisional ability rather than in relation to the choice reached.

The Wisconsin Supreme Court case of *In Re Virgil D* concerned a paranoid chronic schizophrenia sufferer refusing psychotropic medication.[960] The legal issue under consideration was the interpretation of the state statutory definition of incompetence in light of Virgil's refusal.[961] The court interpreted this definition to mean

---

[955] This chimes with the work of sociologists, especially Erving Goffman who have studied the nature of 'total institutions', such as hospitals, schools and prisons. In *Asylums* (1961), Goffman notes: "The point is not that the hospital is a hateful place for patients, but that for the patient to express hatred of it is to give evidence that his place in it is justified and he is not yet ready to leave it" (1961: 335).

[956] Kennedy and Grubb (2000: 764).

[957] See Murphy (1979b) for an argument along these lines.

[958] 148 N.J.Super. 168, 372 A.2d 360 (N.J.Super.Ch. Feb 4, 1977).

[959] *Ibid.* at pp. 180–181 (Dwyer, J.S.C).

[960] 524 N.W.2d 894, 895 (Wis. 1994).

[961] Wisconsin State Alcohol, Drug Abuse, Developmental Disabilities and Mental Health Act, §51.61(1)(g) 4.

a failure to understand the advantages and disadvantages of accepting medication or treatment, and the alternatives to the particular medication or treatment offered.[962] In the view of the court, the definition did not require the patient to believe herself to be mentally ill or to appreciate its nature when refusing psychotropic drugs[963] and, as Berg et al. acknowledge, the patient's denial of her condition was "irrelevant" to the judgment of competence.[964] This is certainly a progressive judgment, and has close parallels with the *Re C* test and the position under PTDCJ. However, it is questionable whether a patient could be judged decisionally competent to consent to treatment or a research procedure to alleviate a condition of which she denies existence, as she would exhibit a failure of belief and appreciation.

Elsewhere, standards of decisional incompetence also ground determinations of FPCLC. In the New Jersey case of *In Re Conroy*,[965] Schreiber J ruled that a determination of incompetence might be made because of a failure to understand information relevant to the decision, to evaluate respective options associated with that decision, or of communication.[966] This ruling is perfectly reasonable, provided sincere attempts have been made on a number of occasions to remedy perceived failures of understanding and to elicit communication skills through education and support. The *Conroy* ruling stands at odds with the more questionable judgment in the Californian case of *Riese v St. Mary's Hospital*,[967] where requirements of understanding, appreciation and evaluation were supplemented by the ability to reach a decision through employing a "rational thought process".[968] The requirement of a "rational thought process" is heavily value-laden and discretionary, and may give rise to an outcome test of competence, particularly if the rationality of the thought process is defined in terms of the competence assessor's view of what constitutes such a process. A "rational reasons" test, alternatively, such as that developed in *US v Charters*[969] (influenced by Benjamin Freedman's account[970]), would be more likely to mitigate against excessive objectively in judging how the reasoning is undertaken and instead look to consistency between the reasoning process and the conclusion reached.[971]

In *Thor v Superior Court*,[972] which involved a ruling on the decisional competence of a quadriplegic prisoner refusing artificial nutrition and medication, the Californian Supreme Court endorsed a similar standard of decisional competence to *Virgil* but with clearer elaboration. Quoting with approval the judgment in

---

[962] 524 N.W.2d 894, 895 (Wis. 1994) at p. 895–896 (Geske J.).

[963] *Ibid.*

[964] Berg et al. (1996: 354, n. 24).

[965] 486 A.2d 1209 (N.J. 1985).

[966] *Ibid.* at p. 1241.

[967] 243 Cal. Rptr 241 (1987).

[968] *Ibid.* at p. 254.

[969] 829 F.2d 479.

[970] 829 F.2d 479 at p. 497 (Murnaghan, J).

[971] Berg et al. (1996: 358, n. 49) also appear to share this view.

[972] 855 P.2d 375 (Cal. 1993).

Rasmussen v Fleming, Arabian J framed the criteria as "the capacity to reason and to make judgments ... a clear understanding of the risks and benefits of the proposed treatment alternatives or nontreatment, along with a full understanding of the nature of the disease and the prognosis".[973] On this basis, the prisoner was found to retain decisional competence to refuse.

The first criterion in *Thor* and *Rasmussen* relating to reasoning and judgment is a central component of any coherent standard of decisional competence. In the context of medical research, however, an awareness of the risks and benefits of alternative treatment options cannot apply. Instead, a measure of understanding the risks involved in participation is appropriate, combined with an appreciation that, in many cases, the benefits of the procedure will accrue to individuals other than the patient or volunteer, possibly not for some time yet.[974] We cannot know, given the absence of case law, which standard would actually be applied. In the context of medical research, it is open to speculation whether a "full understanding of the nature of the disease" refers to the disease that the research intends to investigate, irrespective of whether the patient suffers from this disease or whether the potential participant satisfactorily understands the purpose that the research serves in furthering understanding into that disease.

Appreciation of relevant information is a fundamental component of FPCLC determinations in many states of the US. In *Lane v Candura*,[975] the Massachusetts Appeals Court found that an elderly woman retained decisional competence to refuse potentially life-saving treatment to amputate her gangrenous leg. Psychiatric assessments showed that the respondent could appreciate the "nature and consequences of her act" since she understood that her leg was infected and that death would probably follow from refusing treatment.[976] This contrasts with the ruling of Abrams J in the Massachusetts case of *In Re Roe*,[977] who held that a schizophrenia sufferer's refusal of medication and denial of his mental illness was sufficient to warrant a judgment of incompetence. The consequence of the denial prevented him from appreciating the risks involved in refusing antipsychotic medication, even though he could understand the risks associated with taking it.[978] The decisions in

[973] *Ibid.* at p. 381, originally taken from *Rasmussen v Fleming*, 154 Ariz. 207, 216, 741 P.2d 674, at p. 683; also quoted in Berg et al. (1996: 360).

[974] Berg et al. (1996: 360) consider that the court's reference to "full understanding" may amount to an acknowledgment of a criterion of appreciation in all but name.

[975] 376 N.E.2d 1232 (Mass. App. Ct. 1978).

[976] *Ibid.* at p. 1235–1236.

[977] 583 N.E.2d 1282 (Mass. 1992).

[978] *Ibid.* at p. 1236. See also *Washington v Harper* (1990) 494 U.S. 210, 110 S.Ct. 1028, which clarified the role of the Fourteenth Amendment of the US Constitution (the requirement for "due process of law" in state legal proceedings that may lead to the removal of "life, liberty or property") in the context of a competent refusal of antipsychotic medication. In the words of Kennedy J, the patient "possesses a significant liberty interest in avoiding the unwanted administration of antipsychotic drugs under the Due Process Clause of the Fourteenth Amendment" (*ibid.* at p. 1036) but this right does not require judicial authorization prior to involuntary treatment – the decision could be made by psychiatrists *(ibid.* at p. 1031). McCarron (1990) offers a detailed discussion of refusal of antipsychotic medication and the Constitution.

Lane and *Roe* indicate how treatment refusals may be more readily admitted if they affect or potentially affect only the patient herself but are not admitted in circumstances where denial of illness or refusal to take medication may have an impact both herself and upon other members of society. This highlights the ethically problematic boundary between individual autonomy and legal paternalism in decisional competence determinations.[979]

Recourse to legal paternalism in psychiatric contexts is justified under the PGC only if the effect of interfering with the generic rights of one agent (e.g. liberty) will prevent the probable violation of more important generic rights of other agents (e.g. life). This does not commit us to a risk-relative standard of competence, however, as the appreciation of the risk of harm is subsumed within the criteria for having competence at all, not a function of whether we respect the subsequent choice. On this basis, for example, we might judge an individual who has experienced psychotic episodes of a physically violent nature competent to make a decision about accepting anti-psychotic treatment or some other appropriate treatment, because she recognises the severity of her behaviour and the value of seeking to minimize its harmful effects and the probability of it occurring. This is irrespective of whether she believes herself to be ill in the relevant sense. One may object that this diminishes the salience of psychiatric diagnosis and classification in such circumstances. But the issue relevant to justifying therapeutic legal paternalism under the PGC is not whether a person fails to or chooses not to believe that she is suffering from a recognised psychiatric illness according to a particular nosology. The issue is whether she fails to or chooses not to believe that her behaviour if untreated has the clear potential to violate the rights of other agents. If she does believe this, then it is appropriate for her to become involved in decision-making about appropriate therapeutic interventions that may address the psychological causes of this behaviour. These might include, but are not exhausted by, standard and novel forms of psychiatric treatment.

Reflecting on the myriad legal standards attached to competence across the US, Grisso and Appelbaum have persuasively argued for a definition of incompetence that, while not to be found within law, encapsulates existing legal principles within an explicitly decision-relative standard. Incompetence, they propose:

> constitutes a status of the individual that is defined by functional deficits (due to mental illness, mental retardation or other mental conditions) judged to be sufficiently great that the person currently cannot meet the demands of a specific decision-making situation, weighed in light of its potential consequences.[980]

The reference to the consequences of the decision-making situation does not entail that an individual could be competent to consent to a procedure but not to refuse (or vice-versa), as under an asymmetrical standard of competence. Instead, we can more accurately interpret it as a requirement for an understanding and appreciation of the long-term implications of consenting to or refusing a particular procedure,

---

[979] For a discussion of the problems of paternalism in a medical context, see Buchanan (1978).

[980] Grisso and Appelbaum (1998: 27).

which is central to PTDCJ. This demands a holistic appreciation of the consequences of consent and refusal that recognises competence to make any choice open under the decisional scenario concerned. The definition proposed by Grisso and Appelbaum therefore bears a strong affinity with PTDCJ. Nonetheless, this definition has no legal force, and in its present form is more amenable to influence psychiatric rather than judicial opinion.

Overall, competence to consent to treatment and research in the United States has developed on an *ad hoc* basis. It is still in a state of flux. Certainty and consistency are impeded through the existence of separate standards across states, although it is encouraging to observe that many of them in recent years have tended toward decision-relativity rather than risk-relativity. But for as long as a multitude of state-level tests of competence persist, it is easier to assume that little needs to be done at federal level, which would explain the relative silence upon decisional competence in the FPPHS. The plurality of legal competence assessment standards that exists in the US substantiates Milton Green's observation almost seventy years ago that judicial tests of competence are inherently subjective and lack definition.[981] Although we can see an emergence of a trend towards decision-relativity, it appears that in the intervening years, aside from this, little has changed.

Policy Documents on Decisional Competence

President's Commission: *Making Health Care Decisions* (1982)

We first considered the report *Making Health Care Decisions* briefly at the beginning of Chapter 6. We return to it here to study more closely the discussion of competence it generates. One of the central objectives of this report was to consolidate the recommendations of the earlier Belmont Report (discussed in Chapter 2) by defining more clearly the nature of decisional competence to consent, which they referred to as the non-legal standard of mental capacity.[982] The Commission set out their views in some detail:

> **Elements of capacity:** In the view of the Commission, any determination of the capacity to decide on a course of treatment must relate to the individual abilities of a patient, the requirements of the task at hand, and the consequences likely to follow from the decision ...[983]
>
> **Standards for assessing capacity:** An assessment of an individual's capacity must consider the nature of the particular decision making process in the light of these developments: Does the person possess the ability to understand the relevant facts and alternatives? Is the patient weighing the decision within a framework of values and goals? Is the person able

[981] Green (1941: 145–147), excerpted in White (1994: 10). Roth (1979) and White (1994) have more recently endorsed this view.

[982] President's Commission (1982).

[983] President's Commission (1982: 57).

> to reason and deliberate about this information? Can the patient give reasons for the decision, in the light of the facts, the alternatives and the impact of the decision on the patient's own goals and values? ... Since the assessment must balance competing considerations of well-being and self-determination, *the prudent course is to take into account the potential consequences of a patient's decision* [emphasis added]... the Commission rejects the expressed preference standard for decisions that might compromise the patient's well-being.
> The Commission also rejects as the standard of capacity any test that looks solely to the content of the patient's decision. Any standard based on 'objectively correct' decisions would allow a health professional (or other third party) to declare that a patient lacks decision-making capacity whenever a decision appears 'wrong', 'irrational' or otherwise incompatible with the evaluators view of what is best for the patient.[984]

The standards for assessing capacity expounded by the Commission are generally synonymous with the five general constitutive abilities of decisional competence discussed in Chapter 1. There is a welcome balance inherent in the Commission's proposals through its clear admission that both well-being *and* patient autonomy are relevant considerations in determining competence to consent to research. This meshes closely with PTDCJ. The Commission is also right to be sceptical of the expressed preference standard in cases where the individual's well-being is at stake (such as in biomedical research), since it can veil deeper shortcomings in decisional competence that the requirement to give reasons would expose.

However, the recommendation that the consequences of a decision are relevant to the determination of capacity indicates possible support for a risk-relative (asymmetrical) standard of competence. Aside from their logical incoherence, risk-relative standards are particularly difficult to defend in research settings, where ascribing FPCLC to decide one way but not the other could be seen either as a licence for exploitation or an instance of stringent medical paternalism. If interpreted along risk-relative lines, these proposals would make competence less dependent upon the inherent decisional capabilities of the potential participant than on the choice reached. This is because the acknowledgment of "the potential consequences of a patient's decision" would render competence contingent upon choice-sensitive risk, rather than accounting for risk at the level of the decision itself, as in decision-relative theories. Instead, the level of decisional competence which supported decision-making practices may be able to elicit from any potentially decisionally competent individual should determine the kind of research procedures in which they can meaningfully consent to become involved.

Concerns over how far the Commission may have entertained risk-relativity attenuate somewhat when they go on to identify the outcome, status and functional tests for assessing capacity, endorsing the functional approach. On this basis, the Commission turn to the issue of the decisional competence of individuals suffering from mental disorder:

> [P]atients who are presumed to be incapacitated on the basis of their status may actually be capable of making particular health care decisions ...[M]ildly or moderately retarded individuals hold understandable preferences about healthcare, and the same may be true in

[984] President's Commission (1982: 60–61).

> varying degrees among psychotic persons... Similarly, a senile person may have been declared incompetent by a court and a guardian may have been appointed to manage the person's financial affairs, but the functional standard would not foreclose the need to determine whether the senility also negated the individual's capacity to make health care decisions.[985]

In this paragraph, the Commission appear to endorse a conception of competence that is more closely aligned with PTDCJ. However, the implicit approval given to a risk-relative approach raises the question of which approach is to be preferred when the two conflict.[986] This highlights the intertwined relationship between outcome-dependent risks and risks about which the research procedure requires understanding in order to have competence in the first place. Nevertheless, the explicit recognition that competence to make healthcare decisions may still persist even if competence has diminished in other fields of activity and personal affairs is evidence of the Commission's support for preserving decisional competence as far possible.

Further in the report, the Commission consider the position of children:

> [R]ather than considering children under the age of majority incompetent to decide unless they come within one of the exceptions created by the statutory and common law, these patients could be regarded as competent unless shown to lack decision making capacity.[987]

The proposal for a rebuttable presumption of decisional competence is the antithesis of the rebuttable presumption of decisional incompetence formulated in *Gillick*. In proposing this, the Commission have shown willingness to consider maximising children's autonomy in healthcare decision-making contexts. However, on the evidence considered in Chapter 6, this presumption of decisional competence is probably only appropriate in treatment situations with children above the age of twelve or fourteen and in research situations with children above the age of fourteen or fifteen. Below this, it would seem advisable that a rebuttable presumption of decisional incompetence should apply in both contexts, with higher burden of proof in research contexts.

Expert Panel Report to the National Institutes of Health: *Research Involving Individuals with Questionable Capacity to Consent* (1997)

In 1997, a panel of leading US psychiatrists and neuroscientists were convened by the director of the National Institutes of Health to consider the assessment of decisional competence to consent to medical research on behalf of individuals whose capacity to do so was in doubt.[988] Instead of focusing on the nature and extent of

---

[985] President's Commission (1982: 170).

[986] An example of this would be where an individual has been found competent to consent to a research procedure consisting of several stages, but chooses to consent only to those stages that involve a greater level of risk of harmful side effects.

[987] President's Commission (1982: 170).

[988] Expert Panel Report to the National Institutes of Health (1997a, b).

"shared values" (although areas of consensus were identified),[989] the panel felt that it was more useful to focus on the capacities, impairments, and needs of individuals with questionable ability to consent. Arguing that no "gold standard" exists to assess decisional competence, the Panel concluded:

> ... neither "evidencing a choice" nor correctly answering a given number of factual questions as a measure of "understanding" a protocol are adequate criteria for consent capacity. Instead the subject's *appreciation* [original emphasis] of how the study applies to them and can affect their lives would be preferable criteria. ... A key factor in potential participants' decision-making is their appreciation of how the study applies to them in the context of their lives. Measures of appreciation also link to the potential participant's sense of autonomy, and to the possible coercive elements inherent in his or her situation.[990]

The Panel subsequently comment in the same section:

> While this group focused initially on attempting to identify certain diagnostic groups for whom capacity to consent would be questionable, they concluded that such an approach was flawed. Instead, clinical syndromes or symptoms that might impair attention or memory, communication or motivation, or lead to distractibility or impulsivity were considered more useful to identify.[991]

The report advocates allowing competence or incompetence to be determined by internal characteristics demonstrated by the individuals in question, rather than by their successfulness in meeting a series of predefined generic criteria of understanding or by their belonging to a specific group of individuals whom are traditionally presumed to lack decisional competence. By construing competence in terms of appreciation, the Panel demonstrate their concern to ensure that standards of decisional competence are principally comprised of higher-level cognitive processes. Their emphasis on appreciation may lead to a more clear demarcation between institutionalised individuals whose competence to consent to research is undermined by hopes and fears relating to the consequences of participation for themselves and those for whom the institutional setting apparently has not impacted upon their appreciation of the actual purpose of the research.

Another beneficial insight provided by the Panel is the concentration on specific clinical symptoms or syndromes as being responsible for eroding competence rather than recognised medical conditions. It is possible to distinguish between, for example, a group of depressed patients who have sufficient motivational capabilities to offer a competent consent to a research procedure and those who have not. PTDCJ endorses this, as the judgment is made upon the direct causal relationship between the effect of the symptom and decision-making rather than the more vague relationship between the condition and decision-making. However, a move towards such an approach to understanding impairments of decisional competence would

---

[989] (1997a). In particular, mindful of the fundamental ethical principles enshrined in the Belmont Report, the Panel asserted in the Introduction: "Research must be conducted with integrity and fidelity to the principles of respect for persons, beneficence, and distributive justice. Clinical scientists need to enhance their ethics knowledge base and improve the quality of relationships with research participants" (1997a).

[990] Expert Panel Report to the National Institutes of Health (1997b).

[991] Expert Panel Report to the National Institutes of Health (1997b).

require that professionals responsible for assessing competence be consistently prepared to see beyond generalisations associated with the condition and devote the time necessary to undertake a meaningful assessment of decisional competence.

National Bioethics Advisory Commission: *Research Involving Persons with Mental Disorders That May Affect Decisionmaking Capacity* (1998)

This report, published by the National Bioethics Advisory Commission (NBAC) in 1998,[992] comprises twenty-one recommendations that the Commission believes should govern all research involving such individuals in the United States. Several of these recommendations pertain to competence to consent:

> Informed Consent to Research
> Recommendation 6. No person who has the capacity for consent may be enrolled in a study without his or her informed consent. When potential subjects are capable of making informed decisions about participation, they may accept or decline participation without involvement of any third parties.
> [....]
> *Assessing Potential Subjects' Capacity to Decide about Participating in a Research Protocol*
> Recommendation 8. For research protocols that present greater than minimal risk,[993] an IRB should require that an independent, qualified professional assess the potential subject's capacity to consent. The protocol should describe who will conduct the assessment and the nature of the assessment. An IRB should permit investigators to use less formal procedures to assess potential subjects' capacity if there are good reasons for doing so.
> [....]
> *Expanding Knowledge about Capacity Assessment and Informed Consent*
> Recommendation 19. The National Institutes of Health (NIH) should sponsor research to expand understanding about decisionmaking capacity, the best means for assessing decisionmaking capacity, techniques for enhancing the process of informed consent, and the possible roles of surrogate decision makers in research.[994]

Two issues arise from the NBAC guidance. The provision for the appointment of an external and independent competence assessor prescribed by the research protocol in Recommendation 8 represents a desire to eliminate doubt as far as possible during the informed consent process, and raises the likelihood of an external assessor being better qualified and experienced in making judgments of competence. However, there is no obvious reason why this should apply only to those procedures that involve greater than a minimal risk, since many research procedures that occur, particularly of a therapeutic nature, pose a minimal risk or less to the participant.

---

[992] National Bioethics Advisory Commission (1998).

[993] 'Minimal risk' is defined by US federal legislation as where "the probability and magnitude of harm or discomfort anticipated in the research are not greater in and of themselves than those ordinarily encountered in daily life or during the performance of routine physical or psychological examinations or tests" 45 CFR 46.102(i).

[994] National Bioethics Advisory Commission (1998, Chapter 5).

We also need to clarify the grounds upon which less 'formal' procedures for establishing consent can operate, so that they may be scrutinized properly by IRBs. The exhortation in Recommendation 7 to respect any refusal or objection is a welcome addition, but there is a risk that an investigator could place undue influence on a refusing individual through subsequently approaching her to ascertain whether she has changed her mind. This is more likely when the refusing individual is a patient under the care of the investigator.

The second issue is less encouraging. The absence of any specific criteria with which to make assessments of competence or upon which to judge competence is disappointing. There is also no reference to criteria for decisional or legal competence, or even one propounded by others, in the accompanying commentary. The unqualified use of informed consent compounds this shortcoming.[995] However, the clear misgivings the members of the Commission have about presuming individuals with mental disorders to be decisionally incompetent suggests that NBAC also supports the functional tests for competence such as that endorsed by the President's Commission.

The NBAC recommendations capture the spirit of PTDCJ and are welcome as evidence of a progressive move, but do not go far enough in terms of making clear recommendations for assessing and judging competence. Though the Commission has offered some important procedural requirements for securing and respecting informed consent, the grounds on which this consent should be given and how decision-making can be supported appear to have been passed over. The omission of criteria to establish competence undermines our ability to draw a correspondence between the NBAC guidance and PTDCJ.

NIH *Policy and Guidelines on the Inclusion of Children as Participants in Research Involving Human Subjects* (1998)

Recommendations contained within a Congressional Directive to the NIH in 1995 formed the basis of this NIH Policy.[996] They state that insufficient attention had been focused previously on the ethics of paediatric research, particularly when the potential benefits to children as a whole were considerable. Three years later, this guidance was issued, which makes clear NIH policy in relation to medical research with children. The NIH propose that children should take part within research on human participants which the NIH funds or stages, "unless there are scientific and ethical reasons not to include them". An application for a research protocol which excludes children must provide "an acceptable justification" for this.[997]

The NIH policy is motivated by good intentions to catalyse research projects that have as their aim the furthering of scientific knowledge to inform paediatric medicine.

---

[995] It is likely, however, that the NBAC implicitly understand that informed consent will be construed in accordance with §46.116, Basic DHHS Policy for the Protection of Human Research Subjects, 45 Code of Federal Regulations Part 46.

[996] National Institutes of Health (1998).

[997] National Institutes of Health (1998).

It does, however, display a significant omission in that there is no concept of competence or informed consent contained within these guidelines. It is true that had the NIH devised criteria of competence to consent to research, it would unlikely to have been compatible with the prevailing federal statute, which only contains a concept of assent, as we saw above. Still, it is unfortunate that the NIH policy appears to has been formulated along similarly restrictive lines, rather than spearheading a drive towards new policy.

Whilst it is clear that research with children is necessary in order to discover the aetiology and pathogenesis of disease, this should not be at the expense of giving sufficiently able children the opportunity themselves to offer a competent consent or refusal to research participation. Assent is appropriate for young children, but it is bizarre to apply a blanket policy whereby older adolescents, who are essentially adults for developmental purposes, should be subject to the same principles. The NIH guidance appears something of a hastily conceived reaction to the problem of an insufficient number of child participants in research, whilst maintaining a sufficient range of safeguards. There needs to be a more thoughtful and nuanced approach towards the participation of children of different ages in medical research if these recommendations are to reflect the wide range of their decisional abilities.

Summary

The regulatory status of decisional competence to consent to medical research in the United States remains poorly articulated, in spite of the existence of federal legislation on research and the creation of numerous policy documents to attempt to improve the position. Standards of FPCLC vary between states and are elaborated typically in relation to medical treatment rather than research. There is some uncertainty surrounding how competence is to be assessed and a palpable tension between standards of decisional competence and FPCLC. That the FPPHS does not elaborate a standard of decisional competence or FPCLC undermines the prospects for the development of a consistent standard that would harmonize practice between individual states. There is, however, evidence of some encouraging decision-relative tests of decisional competence in certain states, and Appelbaum and Grisso's definition of incompetence in particular posits a meaningful way forward.

Encouragingly, policy documents suggest that having FPCLC and experiencing cognitive vulnerability are not mutually exclusive. The recommendations of the President's Commission for the Study of Ethical Problems in Medicine and Biomedical and Behavioural Research, the Expert Panel Report to the National Institutes of Health and the National Bioethics Advisory Commission concur that mental disorder should not automatically serve to remove decisional competence in medical contexts, including research. Moreover, case law at state level indicates that FPCLC can be held in conjunction with the existence of a mental disorder or impairment. These judicial pronouncements are clearly a positive development from the perspective of PTDCJ. The NBAC report obliquely refers to the need for further research into supported decision-making, but this appears to be more at the level of aspiration rather than offering an argument for why and how it should be realised. Aside from that, however, there is no evidence of regulatory endorsement of supported decision-making in the US.

In relation to children, the most encouraging ideas are contained in the report of the President's Commission. This study is, however, now over twenty-five years old and further work to identify categories of young people where FPCLC should be presumed, ascribed on a case-by-case basis and not presumed is long overdue. Current federal legislation regulating research with children is grounded in an extensive model of parental legal responsibility and forecloses legal recognition of the decisional competence of many young people by allowing assent only. This position is unjustifiably rigid and the requirement for assent is inappropriate with mature adolescents.

The parity that exists between decisional competence and FPCLC in the US and PTDCJ is largely due to rulings in individual cases and policy documents, rather than because of the FPPHS. Most of the policy documents have explicitly upheld a model of decision-making that is essentially decision-relative and one hopes that it is only a matter of time until it is absorbed through further amendments to the FPPHS into federal law. Overall, however, there is currently insufficient parity between the values of PTDCJ and those embodied in US law. Indeed, in some respects, the regulatory position in the US appears less sympathetic to it than that in England and Wales.

Conclusion

Throughout this book, I have sought to provide an answer to Jeffrie Murphy's question "When in doubt, which way should we err – on the side of safety or on the side of liberty?",[998] in the context of decisional competence to consent to biomedical research. The analysis we have undertaken draws out the explicitly moral nature of the question and places the moral theory grounding the argument at the forefront of this analysis. It has also illustrated the steps that we need to take to move from the abstract dimensions of the theory to its practical implications in judging decisional competence. Recourse to a moral theory is a fundamental precondition of being able to elaborate a clear position in relation to making judgments about competence at all.

For the 'bioethics revolution'[999] to be more than mere rhetoric, it needs to be capable of constantly reassessing, redefining and reanalysing its core assumptions. In the case of decisional competence, conceptual and empirical analysis is warranted to understand how it is displayed, how it may be enhanced and how it may inform (and transform) ethical and legal discourse. Like most bioethical issues, competence cannot and should not be analysed in isolation from the disciplines that bear on it. Decisional competence thus represents an area of bioethics where philosophy, medicine and law meet.[1000] At its core, competence is a normative, value-laden notion whose meaning is formed by the moral theory with which we choose to analyse it.

Applying the PGC has given us a non-arbitrary starting point for pitching judgments of decisional competence and a dialectically necessary basis for our precautionary judgments. The theory of PTDCJ I have derived from it is a decision-relative (yet still risk sensitive) precautionary theory for making judgments about decisional competence. Its reliance upon a foundationalist moral theory and the context of psychiatric studies of decisional competence unites the strengths of theoretical and practical reason in making moral judgments. Precautionary judgments about competence represent a synthesis of the judgments we ought to make about those who are ostensibly agents and the judgments that we can make about their competences, given the

[998] Murphy (1979a: 174), excerpted in full on p. 3 of this book.

[999] Bloche (1998: 45).

[1000] Morreim (1991).

available evidence and the likely impact of the exercise of those competences on the generic rights of themselves and other agents. In proposing the PGC as the basis of my theory, I do not expect individuals suddenly to abandon any normative theory to which they were previously committed, but at least to consider in detail the PGC and its implications for bioethics.

It is true that the precautionary theory I develop is not necessarily connected to biomedical research and could have been developed in relation to some other decision-making context. However, there are good reasons for having chosen biomedical research as the focus of the book. In the early twenty-first century, we appear to be at a stage where the demand for the participation of cognitively vulnerable individuals in medical research is growing rather than receding due to rapid neuroscientific and genetic developments. Understanding when and on what grounds such individuals should be judged to be competent to consent or to refuse for themselves and how we should help them in coming to decisions is an issue which is of concern as much to bioethicists as it is to psychiatrists, other health care professionals and medical lawyers.

Studying the agentive, task and decisional competences of the five cognitively vulnerable groups under consideration in this book has offered a rich picture of their mental skills and abilities, which may often surpass intuitive understandings. Applying the PGC under precaution has shown us that we must treat all members of these groups as agents of some kind, even where (as may occur in extreme cases) the evidence for this is doubtful. That judgments of competence have to be made under the same conditions of uncertainty as ascriptions of agency is even more difficult where the evidence for competence is itself doubtful, as is the case with many cognitively vulnerable participants.

The best evidence for making ascriptions of decisional competence is that which we collect from interactions with the person concerned. However, if we accept PTDCJ, it is not open to us to be neutral about the possibility of competence. Before we can make a morally defensible judgment of competence, we must have satisfied ourselves that we have done all we can to have elicited the individual's latent decisional competence, even if we subsequently find that the individual is incapable of making decisions for herself. Therefore, all those involved in the process of assessing competence and those responsible for making competence judgments both in the medical and legal arenas have a role in ensuring that the available evidence is as thorough as it can be and places the individual's potential for decision-making (however slim it is) in its best possible light.

Judgments of competence under PTDCJ are, then, not mere observations. Because the PGC requires that we assist individuals if they are unable by their own efforts to secure their rights to well-being, there is an obligation as far as possible to support and enhance the decisional abilities of individuals who otherwise may not be able to decide for themselves. This precept informs the account of supported decision-making I propose. Supported decision-making is an important step in making precautionary judgments about any decisional competence. Although I have not sought in this book to propose a particular practical model for going about this, in general terms we can support decision-making by devising and employing methods which are sensitive to the heightened vulnerabilities which affect the potential participant and connect with her way of coming to understand her social world. Only once these methods have been exhausted can we then reach a judgment that this individual is 'incompetent' to make the decision in question and to consider the role of proxy or surrogate decision-making.

Given pervasive conditions of uncertainty in our efforts to gather knowledge, no concept of competence judgment will be flawless. Buchanan has claimed that "[a]ny measurement of capacity is subject to error, and any legal judgment that someone is competent to make a decision that is based on a measurement of capacity will be similarly susceptible."[1001] Buchanan is right that the margin for error in judgments of decisional competence will be reflected in the margin of error exhibited by FPCLC judgments, particularly if the first becomes the necessary and sufficient condition of the second. But this should not cause us to abandon the insights of PTDCJ. Doing something to improve the basis on which competence judgments are ascribed, particularly at the level of first principle, is better than allowing the current unclear and fragmented position to continue unquestioned. I do, however, agree with Buchanan when he suggests later in the same article that "the source and level of doubt attaching to any conclusion regarding capacity should form part of any evidence going to legal competence since it will contribute to the subsequent adjudication of that competence."[1002]

As PTDCJ shows, we can only base judgments of competence on the available evidence in the light of the consequences of believing that evidence for the ostensible agent's generic rights and those of any other agents to be affected. Where a legal decision requires a decisive answer, we must always err on the side of that judgment which is the least restrictive of the generic rights of the agents involved, following the criterion of avoidance of more probable harm. The most appropriate and least confusing way to resolve this is to make a decision-relative account of decisional competence the necessary and sufficient condition of FPCLC. This would remove the possibility of conflation with other types of legal competence, and even with mental capacity.

At the end of his article, 'Competence, Marginal and Otherwise', Benjamin Freedman states: "An important task for philosophy, particularly in the medical/legal interface, is the exposure of muddled and wrong-headed concepts, to clear the way for a healthy growth of ideas".[1003] Decisional competence, in particular, has been beset by these confusions. I hope to have at least exposed the source of these 'muddles' and laid the foundations for further research in this area. The importance of Murphy's original question which inspired this book has not diminished in the intervening years; indeed, the rate of change in biomedical research has rendered it all the more urgent. If the argument of this book is accepted, then we have arrived at a defensible theoretical framework for making judgments of decisional competence. I commend it to competence assessors and policy makers as a rational and compassionate way to secure empowerment and protection for cognitively vulnerable people.

[1001] Buchanan (2004: 417).

[1002] Buchanan (2004: 417–418).

[1003] Freedman (1981: 72).

References

Abernethy, V. (1984). 'Compassion, control, and decisions about competency'. *American Journal of Psychiatry*, vol. 141, pp. 53–60.
Abernethy, V. (1991). 'Judgments about Patient Competence: Cultural and Economic Antecedents'. In: M. A. G. Cutter and E. E. Shelp (eds.) *Competency: A Study of Informal Competency Determinations in Primary Care*, Kluwer Academic, Dordrecht, pp. 211–226.
Abernethy, V. and Lundin, K. (1980). 'Patient Competence: Conditions for Giving or Withholding Consent'. In: V. Abernethy (ed.) *Frontiers in Medical Ethics*, Ballinger, Cambridge, pp. 79–98.
Abrams, N. (1977). 'Medical Experimentation: The Consent of Prisoners and Children'. In: S. F. Spicker and H. Tristam Engelhardt Jr. (eds.) *Philosophical Medical Ethics: Its Nature and Significance*, D. Reidel, Dordrecht, pp. 111–124.
Ackerman, S. J. (2006). *Hard Science, Hard Choices: Facts, Ethics and Policies Guiding Brain Science Today*, Dana, New York.
Adamis, D., Martin, F C., Treloar, A. and Macdonald, A. J. D. (2005). 'Capacity, consent, and selection bias in a study of delirium'. *Journal of Medical Ethics*, vol. 31, pp. 137–145.
Adults with Incapacity (Scotland) Act 2000 (2000 asp 4). Available at: http://www.opsi.gov.uk/legislation/scotland/acts2000/asp_20000004_en.pdf [Accessed 30/03/08].
Advisory Committee on Human Radiation Experiments (1996). *The Human Radiation Experiments*, Oxford University Press, New York.
Alderson, P. (1992). 'In the Genes or in the Stars? Children's Competence to Consent'. *Journal of Medical Ethics*, vol. 18(3), pp. 119–124.
Alderson, P. (1993). *Children's Consent to Surgery*, Open University Press, Buckingham.
Alderson, P. and Montgomery, J. (1996). *Health Care Choices: Making Decisions with Children*, Institute for Public Policy Research, London.
Allebeck, P., Varla, A. and Wistedt, B. (1986). 'Suicide and Violent Death Among Patients with Schizophrenia'. *Acta Psychiatrica Scandinavia*, vol. 74(1), pp. 43–49.
Alora, A. T. and Lumitao, J. M. (eds.) (2001). *Beyond a Western Bioethics: Voices from the Developing World*, Georgetown University Press, Washington, DC.
Alzheimer Europe (undated). *Position Paper on Research with People with Dementia: Executive Summary*. Available at: http://www.alzheimer-europe.org/index.php?lm2 = 225A8956AC31 [Accessed 30/03/08].
Alzheimer Europe (2006). *The Population of People with Dementia in Europe*. Available at: http://www.dementia-in-europe.eu/?lm2 = OWQAUJKRXAEZ [Accessed 30/03/08].
Alzheimer's Society (2007). *Dementia UK: The Full Report*. Alzheimer's Society, London. Available at: http://www.alzheimers.org.uk/downloads/Dementia_UK_Full_Report.pdf [Accessed 30/03/08].
American Psychiatric Association (1994). *Diagnostic and Statistical Manual of Mental Disorders* (DSM-IV), fourth edition, American Psychiatric Association, Washington, DC.

Anderson, J. and Honneth, A. (2005). 'Autonomy, Vulnerability, Recognition and Justice'. In: J. Christman and J. Anderson (eds.) *Autonomy and the Challenges to Liberalism: New Essays*, Cambridge University Press, Cambridge, pp. 127–149.

Andreasen, N. C. (2001). *Brave New Brain: Conquering Mental Illness in the Era of the Genome*, Oxford University Press, New York.

Andreasen, N. C. (ed.) (2005). *Research Advances in Genetics and Genomics: Implications for Psychiatry*, American Psychiatric Press, Arlington, VA.

Annas, G. J. (1999). *Some Choice: Law, Medicine and the Market*, Oxford University Press, Oxford.

Annas, G. J. and Grodin, M. A. (eds.) (1995). *The Nazi Doctors and the Nuremberg Code: Human Rights in Human Experimentation*, Oxford University Press, New York.

Appelbaum, P. S. (1997). 'Rethinking the Conduct of Psychiatric Research'. *Archives of General Psychiatry*, vol. 54, pp. 117–120.

Appelbaum, P. S. (1998a). 'Ought We to Require Emotional Capacity as Part of Decisional Competence?' *Kennedy Institute of Ethics Journal*, vol. 8(4), pp. 377–387.

Appelbaum, P. S. (1998b). 'Missing the Boat: Competence and Consent in Psychiatric Research'. *American Journal of Psychiatry*, vol. 155(11), pp. 1486–1488.

Appelbaum, P. S. (2002). 'Involving Decisionally Impaired Subjects in Research: The Need for Legislation'. *American Journal of Geriatric Psychiatry*, vol. 10(2), pp. 120–125.

Appelbaum, P. S. and Grisso, T. (1995). 'The MacArthur Treatment Competence Study: I', *Law and Human Behavior*, vol. 19(2), pp. 105–126.

Appelbaum, P. S. and Grisso, T. (2001). *MacCAT-CR: MacArthur Competence Assessment Tool for Clinical Research*, Professional Resource Press, Sarasota, FL.

Appelbaum, P. S. and Roth, L. H. (1981). 'Clinical Issues in the Assessment of Competency'. *American Journal of Psychiatry*, vol. 138, pp. 1462–1467.

Appelbaum, P. S., Roth, L. and Lidz, C. (1982). 'The Therapeutic Misconception: Informed Consent in Psychiatric Research'. *International Journal of Law and Psychiatry*, vol. 5(3–4), pp. 319–329.

Appelbaum, P. S., Grisso, T, Frank, E., O'Donnell, S., Kupfer, D. J. (1999). 'Competence of Depressed Patients for Consent to Research'. *American Journal of Psychiatry*, vol. 156(9), pp. 1380–1384.

Arboleda-Florez, J. and Weisstub, D. N. (1998). 'Ethical Research with Vulnerable Populations: The Mentally Disordered'. In: D. N. Weisstub (ed.) *Research on Human Subjects: Ethics, Law and Social Policy*, Pergamon, Oxford, pp. 433–450.

Arscott, K., Dagnan, D. and Stenfert Kroese, B. (1999). 'Assessing the Ability of People with a Learning Disability to Give Informed Consent to Treatment'. *Psychological Medicine,* vol. 29, pp. 1367–1375.

Artiga, M. J., Bullido, M. J., Frank, A., Sastre, I., Recuero, M., Garcia, M. A., Lendon, C. L., Han, S. W., Morris, J.C., Vazquez, J., Goate, A. and Valdivieso, F. (1998). 'Risk for Alzheimer's Disease Correlates with Transcriptional Activity of the APOE Gene'. *Human Molecular Genetics*, vol. 7, pp. 1887–1892.

Ashcroft, R., Goodenough, T., Williamson E. and Kent, J. (2003). 'Children's Consent to Research Participation: Social Context and Personal Experience Invalidate Fixed Cut-off Rules'. *American Journal of Bioethics*, vol. 3(4), pp. 16–17.

Ashcroft, R. E., Dawson, A., Draper, H. and McMillan, J. R. (eds.) (2007). *Principles of Health Care Ethics*, second edition, Wiley, Chichester.

Astington, J. W. (1993). *The Child's Discovery of the Mind*, Harvard University Press, Cambridge, MA.

Aune, B. (1961). 'The Problem of Other Minds'. *The Philosophical Review*, vol. 70(3), pp. 320–329.

Bartlett, H. and Martin, W. (2002). 'Ethical Issues in Dementia Care Research'. In: H. Wilkinson (ed.) *The Perspectives of People with Dementia: Research Methods and Motivations*, Jessica Kingsley, London, pp. 47–61.

Basic DHHS (Department of Health and Human Services) *Policy for the Protection of Human Research Subjects, 45 Code of Federal Regulations Part 46*: Revised June 23 2005; Effective

June 23 2005. Available at: http://www.hhs.gov/ohrp/documents/OHRPRegulations.pdf [Accessed 30/03/08].

Bauer, L. (1990). *The Development of the Understanding of Diabetes in Children with Diabetes.* PhD thesis, The University of Toronto, Toronto.

Baum, M. (1990). 'The Ethics of Clinical Research'. In: P. Byrne (ed.) *Ethics and Law in Health Care and Research*, Wiley, Chichester, pp. 1–7.

Baumgarten, E. (1980). 'The Concept of Competence in Medical Ethics'. *Journal of Medical Ethics,* vol. 6, pp. 180–184.

BBC News Online (14/11/01) '*Missing gene increases CJD risk*', Available at: http://news.bbc.co.uk/1/hi/health/1654178.stm. [Accessed 30/03/08].

Beauchamp, T. L. (1991). 'Competence'. In: M. A. G. Cutter and E. E. Shelp (eds.) *Competency: A Study of Informal Competency Determinations in Primary Care*, Kluwer Academic, Dordrecht, pp. 49–77.

Beauchamp, T. L. and Childress, J. F. (2001). *Principles of Biomedical Ethics*, fifth edition, Oxford University Press, New York.

Bech, P. (2000). 'Clinical Features of Mood Disorders and Mania'. In: M. G. Gelder, J. J. López-Ibor Jr. and N. C. Andreasen (eds.) *New Oxford Textbook of Psychiatry: Volume 1*, Oxford University Press, Oxford, pp. 682–688.

Becker L. C. and Becker, C. B. (eds.) (2001). *Encyclopaedia of Ethics*, second edition, Routledge, New York.

Beecher, H. K. (1966). 'Ethics and Clinical Research'. *New England Journal of Medicine,* vol. 274, pp. 1354–1360.

Benaroyo, L. and Widdershoven, G. (2004). 'Competence in Mental Health Care: A Hermeneutic Perspective'. *Health Care Analysis*, vol. 12(4), pp. 295–306.

Benatar, S. R. (2001).'Justice and Medical Research: A Global Perspective'. *Bioethics* vol. 15(4), pp. 333–340.

Berg, J. W. (1996). 'Legal and Ethical Complexities of Consent with Cognitively Impaired Research Subjects: Proposed Guidelines'. *Journal of Law, Medicine and Ethics* vol. 24(1), pp. 18–35.

Berg, J. W. and Appelbaum, P. S. (1999). 'Subjects' Capacity to Consent to Neurobiological Research'. In: H. A. Pincus, J. A. Lieberman and S. Ferris (eds.) *Ethics in Psychiatric Research: A Resource Manual for Human Subjects Protection*, American Psychiatric Association, Washington, DC, pp. 81–106.

Berg, J. W., Appelbaum, P. S. and Grisso, T. (1996). 'Constructing Competence: Formulating Standards of Legal Competence To Make Medical Decisions'. *Rutgers Law Review*, vol. 48(2), pp. 345–396.

Berg, J. W., Appelbaum, P. S., Lidz, C. W. and Parker, L. S. (2001). *Informed Consent: Legal Theory and Clinical Practice*, second edition, Oxford University Press, New York.

Berghmans, R. (1998). 'Advance Directives for Non-Therapeutic Dementia Research: Some Ethical and Policy Considerations'. *Journal of Medical Ethics*, vol. 24, pp. 32–37.

Beyleveld, D. (1991). *The Dialectical Necessity of Morality: An Analysis and Defense of Alan Gewirth's Argument to the Principle of Generic Consistency*, The University of Chicago Press, Chicago, IL.

Beyleveld, D. (1996). 'Legal Theory and Dialectically Contingent Justifications for the Principle of Generic Consistency'. *Ratio Juris*, vol. 9(1), pp. 15–41.

Beyleveld, D. (2000). 'The Moral Status of the Human Embryo and Foetus'. In: H. Haker and D. Beyleveld (eds.) *The Ethics of Genetics in Human Procreation*, Ashgate, Aldershot, pp. 59–85.

Beyleveld, D. (2001). 'Reconstructing Human Dignity by Integrating Vulnerability with Autonomy'. Unpublished paper given at the conference 'Human Rights and Human Dignity in Relation to Public Health and Bioethics', Sheffield Institute of Biotechnological Law and Ethics, Department of Law, The University of Sheffield, 18 July 2001.

Beyleveld, D. (2002). 'A Reply to Marcus G. Singer on Gewirth, Beyleveld and Dialectical Necessity'. *Ratio Juris,* vol. 15(4), pp. 458–473.

Beyleveld, D. (2008). 'Human Cognitive Vulnerability and the Moral Status of the Human Embryo and Foetus'. In: Düwell, M., Rehmann-Sutter, C, and Mieth, D. (eds.) *The Contingent Nature of Life: Bioethics and the Limits of Human Existence*. Springer, Dordrecht, pp. 83–88.

Beyleveld, D. and Brownsword, R. (1986). *Law as a Moral Judgment*, Sweet and Maxwell, London.
Beyleveld, D. and Brownsword, R. (2001). *Human Dignity in Bioethics and Biolaw*, Oxford University Press, Oxford.
Beyleveld, D. and Brownsword, R. (2006). 'Principle, Proceduralism and Precaution in a Community of Rights'. *Ratio Juris*, vol. 19(2), pp. 141–168.
Beyleveld, D. and Brownsword, R. (2007a). *Consent in The Law*, Hart, Oxford.
Beyleveld, D. and Brownsword, R. (2007b). 'Uses and Abuses of Precautionary Reasoning in Law'. Paper given at the Society of Legal Scholars 2007 annual conference, Durham University, 13 September 2007, on file with the author.
Beyleveld, D. and Pattinson, S. (1998). 'Proportionality, Precaution and Duties in Relation to "PPAs-in-Part"'. Unpublished paper, on file with the author.
Beyleveld, D. and Pattinson, S. (2000). 'Precautionary Reasoning as a Link to Moral Action'. In: M. Boylan (ed.), *Medical Ethics*, Prentice-Hall, Upper Saddle River, NJ, pp. 39–53.
Bielby, P. (2005a). 'Equality and Vulnerability in Biomedical Research on Human Subjects'. *Imprints: Egalitarian Theory and Practice*, vol. 8(3), pp. 219–248.
Bielby, P. (2005b). 'The Conflation of Competence and Capacity in English Medical Law: A Philosophical Critique'. *Medicine, Health Care and Philosophy*, vol. 8(3), pp. 357–369.
Bielby, P. (2008, forthcoming). 'Towards Supported Decision-Making in Biomedical Research with Cognitively Vulnerable Adults'. In: O. P. Corrigan, K. M. Liddell, J. McMillan, M. P. M. Richards and C. Weijer (eds.) *The Limits of Consent: A Socio-Ethical Approach to Human Subject Research in Medicine*, Oxford University Press, Oxford.
Birchwood, M. and Jackson, C. (2001). *Schizophrenia*, Psychology, Hove.
Bloche, M. G. (1998). 'Beyond Consent'. In: D. N. Weisstub (ed.) *Research on Human Subjects: Ethics, Law and Social Policy*, Pergamon, Oxford, pp. 44–55.
Block, N. and Dworkin, G. (eds.) (1977). *The IQ Controversy: Critical Readings*, Quartet Books, London.
Bonnie, R. J. (1997). 'Research with Cognitively Impaired Subjects: Unfinished Business in the Regulation of Human Research'. *Archives of General Psychiatry*, vol. 54(2), pp. 105–111.
Boucher, D. and Kelly, P. (eds.) (1994). *The Social Contract from Hobbes to Rawls*, Routledge, London.
Boylan, M. (ed.) (1999). *Gewirth: Critical Essays on Action, Rationality and Community*. Rowman & Littlefield, Lanham, MD.
Boylan, M. (2000). *Basic Ethics*, Prentice-Hall, Upper Saddle River, NJ.
Braak, E. and Braak, H. (1995). 'Staging of Alzheimer's Disease-Related Neurofibrillary Changes'. *Neurobiology of Aging*, vol. 16(3), pp. 271–278.
Brazier, M. and Lobjoit, M. (eds.) (1991). *Protecting the Vulnerable: Autonomy and Consent in Health Care*, Routledge, London.
Breden, T. M. and Vollmann, J. (2004). 'The Cognitive Based Approach of Capacity Assessment in Psychiatry: A Philosophical Critique of the MacCAT-T'. *Health Care Analysis*, vol. 12(4), pp. 273–283.
Breier, A. and Strauss, J. S. (1983). 'Self-Control in Psychotic Disorders'. *Archives of General Psychiatry*, vol. 40, pp. 1141–1145.
British Medical Association (2001). *Consent, Rights and Choices in Health Care for Children and Young People*, BMJ Books, London.
British Medical Association and The Law Society (2004). *Assessment of Mental Capacity: Guidance for Doctors and Lawyers*, second edition, BMJ Books, London.
Brock, D. W. (1991). 'Decision-Making Competence and Risk'. *Bioethics*, vol. 5(2), pp. 105–112.
Brock, D. W. (1993). *Life and Death: Philosophical Essays in Biomedical Ethics*, Cambridge University Press, Cambridge.
Brody, B. A. (1998). *The Ethics of Biomedical Research: An International Perspective*, Oxford University Press, New York.
Brown, S. P. (1998). *The Moral Justification of Retributive Punishment by Reference to the Notion of Balance*. Unpublished PhD thesis, Department of Law, The University of Sheffield.

Brownsword, R. (2004). 'The Cult of Consent: Fixation and Fallacy'. *Kings College Law Journal*, vol. 15, pp. 223–251.

Brunello, N., Langer, S. Z., Racagni, G. (eds.) (1997). *Mental Disorders in the Elderly: New Therapeutic Approaches (International Academy for Biomedical and Drug Research, vol. 13).* Karger, Basel, Switzerland.

Buchanan, A. (2004). 'Mental Capacity, Legal Competence and Consent to Treatment'. *Journal of the Royal Society of Medicine*, vol. 97, pp. 415–420.

Buchanan, A. E. (1978). 'Medical Paternalism'. *Philosophy and Public Affairs,* vol. 7, pp. 370–390.

Buchanan, A. E. and Brock, D. W. (1990). *Deciding for Others: The Ethics of Surrogate Decision-Making*, Cambridge University Press, New York.

Buckles, V. D., Powlishta, K. K., Palmer, J. L, Coats, M.,. Hosto, T, Buckley, A. and Morris, J. C. (2003). 'Understanding of Informed Consent by Demented Individuals'. *Neurology*, vol. 61, pp. 1662–1666.

Buller, T. (2001). 'Competence and Risk-Relativity'. *Bioethics*, vol. 15(2), pp. 94–109.

Bursztajn, H. J., Harding, H. P., Gutheil, T. G. and Brodsky, A. (1991). 'Beyond Cognition: The Role of Disordered Affective States in Impairing Competence to Consent to Treatment'. *Bulletin of the American Academy of Psychiatry and Law*, vol. 19(4), pp. 383–388.

Cage, J. (1961). *Silence: Lectures and Writings*, Wesleyan University Press, Middletown, CT.

Cale, G. S. (1999). 'Risk-Related Standards of Competence: Continuing the Debate Over Risk-Related Standards of Competence'. *Bioethics* vol. 13(2), pp. 131–148.

Cameron, L. and Murphy, J. (2006). 'Obtaining Consent to Participate in Research: The Issues Involved in Including People with a Range of Learning and Communication Disabilities'. *British Journal of Learning Disabilities*, vol. 35, pp. 113–120.

Candilis, P. J. (2003). 'Early Intervention in Schizophrenia: Three Frameworks for Guiding Ethical Inquiry'. *Psychopharmacology*, vol. 171, pp. 75–80.

Canterbury v Spence 464 F 2d 772 (DC, 1972).

Cantor, N. L. (2005). *Making Medical Decisions for the Profoundly Mentally Disabled*, MIT, Cambridge, MA.

Capron, A. M. (1999). 'Ethical and Human-Rights Issues in Research on Mental Disorders That May Affect Decision-Making Capacity'. *The New England Journal of Medicine*, vol. 348(18), pp. 1430–1434.

Carpenter, W. T., Gold, J. M., Lahti, A. C., et al. (2000). 'Decisional Capacity for Informed Consent in Schizophrenia Research'. *Archives of General Psychiatry*, vol. 57(6), pp. 533–538.

Carruthers, P. (1986). *Introducing Persons: Theories and Arguments in the Philosophy of Mind*, Routledge, London.

Cavadino, M. (1997). 'A Vindication of the Rights of Psychiatric Patients'. *Journal of Law and Society*, vol. 24(2), pp. 235–251.

Chadwick, P., Birchwood, M. and Trower, P. (1996). *Cognitive Therapy for Delusions, Voices and Paranoia*, Wiley, Chester.

Chamberlin, J. (1977). *On Our Own: Patient Controlled Alternatives to the Mental Health System*, McGraw-Hill, New York.

Charland, L. C. (1998a). 'Is Mr. Spock Mentally Competent? Competence to Consent and Emotion'. *Philosophy, Psychiatry and Psychology*, vol. 5(1), pp. 67–81.

Charland, L. C. (1998b). 'Appreciation and Emotion: Theoretical Reflections on the MacArthur Treatment Competence Study'. *Kennedy Institute of Ethics Journal*, vol. 8, pp. 359–376.

Charland, L. C. (2001). 'Mental Competence and Value: The Problem of Normativity in the Assessment of Decision-Making Capacity'. *Psychiatry, Psychology and Law*, vol. 8(2), pp. 135–145.

Charlesworth, M. J. (1993). *Bioethics in a Liberal Society*, Cambridge University Press, Cambridge.

Checkland, D. (2001). 'On Risk and Decisional Capacity'. *Journal of Medicine and Philosophy*, vol. 26(1), pp. 35–59.

Children Act 1989 (c. 41). HMSO, London.

Childress, J. F. (1990). 'The Place of Autonomy in Bioethics'. *Hastings Center Report*, vol. 20, pp. 12–16.

Childress, J. F. (1995). *Who Should Decide? Paternalism in Health Care*, Oxford University Press, New York.

Clucas, B. R. (2004). *The Rights of Children*. Unpublished PhD thesis, Department of Law, The University of Sheffield.

Colman, A. M. (2001). *A Dictionary of Psychology*, Oxford University Press, Oxford.

Cohen, D. and Eisdorfer, C. (1986). *The Loss of Self: A Family Resource for the Care of Alzheimer's Disease and Related Disorders*, W. W. Norton and Co., New York.

Corrigan, O. P. (2003). 'Empty Ethics: The Problem with Informed Consent'. *Sociology of Health and Illness*, vol. 25(3), pp. 768–792.

Council for International Organizations of Medical Sciences (2002). *International Ethical Guidelines for Biomedical Research Involving Human Subjects*, 2002 revision, Council for International Organizations of Medical Sciences (CIOMS) and the World Health Organization (WHO), Geneva. Available at: http://www.cioms.ch/frame_guidelines_nov_2002.htm [Accessed 30/03/08].

Council of Europe Committee of Ministers (1983). *Recommendation No. R 83(2) Concerning the Legal Protection of Persons Suffering from Mental Disorder Placed as Involuntary Patients Without any Discernible Justification*. Adopted by the Committee of Ministers on 22 February 1983. In: Directorate General I – Legal Affairs Bioethics Department (ed.) (2005). *Texts of the Council of Europe on Bioethical Matters: Volume 1*, Council of Europe, Strasbourg, pp. 17–20. Available at: http://www.coe.int/t/e/legal_affairs/legal_co-operation/bioethics/texts_and_documents/INF_2005_2e%20vol%20I%20textes%20CoE%20bioethique.pdf [Accessed 30/03/08].

Council of Europe (1997). *Convention for the Protection of Human Rights and Human Dignity with Regard to the Application of Biology and Medicine: Convention on Human Rights and Biomedicine*, ETS 164. In: Directorate General I – Legal Affairs Bioethics Department (ed.) (2005). *Texts of the Council of Europe on Bioethical Matters: Volume 1*, Council of Europe, Strasbourg, pp. 146–156. Available at: http://www.coe.int/t/e/legal_affairs/legal_co-operation/bioethics/texts_and_documents/INF_2005_2e%20vol%20I%20textes%20CoE%20bioethique.pdf [Accessed 30/03/08].

Council of Europe (2000). White paper on the protection of the human rights and dignity of people suffering from mental disorder, especially those placed as involuntary patients in a psychiatric establishment. Council of Europe, Strasbourg.

Council of Europe (2005). *Additional Protocol to the Convention on Human Rights and Biomedicine Concerning Biomedical Research,* Treaty Series No. 19, Council of Europe, Strasbourg. In: Directorate General I – Legal Affairs Bioethics Department (ed.) (2005). *Texts of the Council of Europe on Bioethical Matters: Volume 1*, Council of Europe, Strasbourg, pp. 174–186. Available at: http://www.coe.int/t/e/legal_affairs/legal_co-operation/bioethics/texts_and_documents/INF_2005_2e%20vol%20I%20textes%20CoE%20bioethique.pdf [Accessed 30/03/08].

Council of Europe (2007). *Convention on Human Rights and Biomedicine: Chart of Signatures and Ratifications,* status on 1 March 2008. Available at: http://conventions.coe.int/Treaty/Commun/ChercheSig.asp?NT = 164&CM = 1&DF = 01/03/2008&CL = ENG.

Council of Europe Committee of Ministers (1999). *Recommendation No. R (99) 4 footnote 1 of the Committee of Ministers to Member States on Principles Concerning the Legal Protection of Incapable Adults*. Adopted by the Committee of Ministers on 23 February 1999. In: Directorate General I – Legal Affairs Bioethics Department (ed.) (2005).*Texts of the Council of Europe on Bioethical Matters: Volume 1*, Council of Europe, Strasbourg, pp. 86–93. Available at: http://www.coe.int/t/e/legal_affairs/legal_co-operation/bioethics/texts_and_documents/INF_2005_2e%20vol%20I%20textes%20CoE%20bioethique.pdf [Accessed 30/03/08].

Culver, C. M. and Gert, B. (1982). 'Valid Consent and Competence'. In: C. M. Culver and B. Gert (eds.) *Philosophy in Medicine: Conceptual and Ethical Issues in Medicine and Psychiatry*, Oxford University Press, New York, pp. 42–63.

Culver, C. M. and Gert, B. (1990). 'The Inadequacy of Incompetence'. *The Milbank Quarterly*, vol. 68(4), pp. 619–643.

Culver, C. M. and Gert, B. (2004). 'Competence'. In: J. Radden (ed.) *The Philosophy of Psychiatry: A Companion*, Oxford University Press, Oxford, pp. 258–270.

Cutter, M. A. G. (1991). 'Introduction'. In: M. A. G. Cutter and E. E. Shelp (eds.) *Competency: A Study of Informal Competency Determinations in Primary Care*, Kluwer Academic, Dordrecht, pp. ix–xviii.

Cutter, M. A. G. and Shelp, E. E. (eds.) (1991). *Competency: A Study of Informal Competency Determinations in Primary Care*, Kluwer Academic, Dordrecht.

Damasio, A. (2005) [1994]. *Descartes' Error: Emotion, Reason and the Human Brain* (reissue with new preface), Penguin, New York.

Dana Alliance for Brain Initiatives (2007). *2007 Progress Report on Brain Research*, Dana, New York.

Dekkers, W. J. M. (2001). 'Autonomy and Dependence: Chronic Physical Illness and Decision-Making Capacity'. *Medicine, Health Care and Philosophy*, vol. 4, pp. 185–192.

DeLong, M. R. and Wichmann, T. (2001). 'Deep Brain Stimulation for Parkinson's Disease'. *Annals of Neurology*, vol. 49(2), pp. 142–143.

DeMarco, J. P. (2002). 'Competence and Paternalism'. *Bioethics*, vol. 16(3), pp. 231–245.

Department of Constitutional Affairs (2007). *Mental Capacity Act 2005 Code of Practice*, TSO, London. Available at: http://www.justice.gov.uk/docs/mca-cp.pdf [Accessed 30/03/08].

Department of Constitutional Affairs and UK Department of Health (2005). *Explanatory Notes to the Mental Capacity Act 2005*, TSO, London.

DeRenzo, E. G. (1997). 'Decisionally Impaired Persons in Research: Revising the Proposed Instruments'. *Journal of Law, Medicine and Ethics*, vol. 25, pp. 139–149.

DeRenzo, E. G., Conley, R. R. and Love, R. (1998). 'Assessment of Capacity to Give Consent to Research Participation: State-of-the-Art and Beyond'. *Journal of Health Care Law and Policy*, vol. 1, pp. 66–87.

Devereux, J. A. (1993). *Competence to Consent to Medical Treatment in England and Australia*. Unpublished D.Phil. thesis, The University of Oxford.

Directive 2001/20/EC of the European Parliament and of the Council on the approximation of the laws, regulations and administrative provisions of the Member States relating to the implementation of good clinical practice in the conduct of clinical trials on medicinal products for human use. *Official Journal of the European Communities*, 01/05/01, L 121/34. Available at: http://europa.eu/eur-lex/pri/en/oj/dat/2001/l_121/l_12120010501en00340044.pdf [Accessed 30/03/08].

Doyal, L. and Tobias, J. S. (eds.) (2000). *Informed Consent in Medical Research*, BMJ Books, London.

Drane, J. F. (1984). 'Competency to Give Informed Consent'. *Journal of the American Medical Association*, vol. 252, pp. 925–927.

Drane, J. F. (1985). 'The Many Faces of Competency'. *Hastings Center Report*, vol. 15, pp. 17–21.

Dresser, R. (1995). 'Dworkin on Dementia: Elegant Theory, Questionable Policy'. *Hastings Center Report*, vol. 25(6), pp. 32–38.

Dresser, R. (1996). 'Mentally Disabled Research Subjects: The Enduring Policy Issues'. *Journal of the American Medical Association*, vol. 276(1), pp. 67–72.

DuBois, J. M. (2007). *Ethics in Mental Health Research: Principles, Guidance, and Cases*, Oxford University Press, New York.

Dunn. L. B. (2006). 'Capacity to Consent to Research in Schizophrenia: The Expanding Evidence Base'. *Behavioral Sciences and the Law*, vol. 24(4), pp. 431–445.

Dunn, L. B. and Jeste, D. V. (2001). 'Enhancing Informed Consent for Research and Treatment'. *Neuropsychopharmacology*, vol. 24, pp. 595–607.

Dunn, L. B., Lindamer, L. A., Palmer, B. W., Schneiderman, L. J. and Jeste, D. V. (2001). 'Enhancing Comprehension of Consent for Research in Older Patients with Psychosis:

A Randomized Study of a Novel Educational Strategy'. *American Journal of Psychiatry*, vol. 158, pp. 1911–1913.
Dworkin, G. (1988). *The Theory and Practice of Autonomy*, Cambridge University Press, Cambridge.
Dworkin, G. (1989). 'The Concept of Autonomy'. In: J. Christman (ed.) *The Inner Citadel: Essays on Individual Autonomy*, Oxford University Press, New York, pp. 54–62.
Dworkin, R. (1978). *Taking Rights Seriously*, Duckworth, London.
Dworkin, R. (1984). 'Rights As Trumps'. In: J. Waldron (ed.) *Theories of Rights*, Oxford University Press, New York, pp. 153–167.
Dworkin, R. (1986). 'Autonomy and the Demented Self'. *The Milbank Quarterly*, vol. 64(2), pp. 4–16.
Dworkin, R. (1994). *Life's Dominion: An Argument About Abortion, Euthanasia and Individual Freedom* (reissued with a new preface), Vintage Books, New York.
Dworkin, R. (2004). 'Replies to My Critics'. In: J. Burley (ed.) *Dworkin and His Critics*, Blackwell, Oxford, pp. 339–395.
Eastman, N. (1992). 'Psychiatric, Psychological and Legal Models of Man'. *International Journal of Law and Psychiatry*, vol. 15, pp. 157–169.
Ebert, D. and Ebmeier, K.P. (eds.) (1998). *New Models for Depression (Advances in Biological Psychiatry, vol. 19).* Karger, Basel, Switzerland.
Edmundson, W. E. (2004). *An Introduction to Rights*. Cambridge University Press, Cambridge.
Elliott, C. (1997). 'Caring About Risks: Are Severely Depressed Patients Competent to Consent to Research?' *Archives of General Psychiatry*, vol. 54(2), pp. 113–116.
Elliott, C. (1998). 'Patients Doubtfully Capable or Incapable of Consent'. In: H. Kuhse and P. Singer (eds.) *A Companion to Bioethics*, Blackwell, Oxford, pp. 452–462.
Erde, E. L. (1999). 'Paradigms and Personhood: A Deepening of the Dilemmas in Ethics and Medical Ethics'. *Theoretical Medicine and Bioethics*, vol. 20, pp. 141–160.
Expert Panel Report to the National Institutes of Health (1997a). *Research Involving Individuals with Questionable Capacity to Consent: Ethical Issues and Practical Considerations for Institutional Review Boards - Introduction*, National Institutes of Health Bioethics Interest Group, Rockville, Maryland. Available at: http://www.nih.gov/sigs/bioethics/reports/intro.htm [Accessed 04/09/02 – No longer available].
Expert Panel Report to the National Institutes of Health (1997b). *Research Involving Individuals with Questionable Capacity to Consent: Ethical Issues and Practical Considerations for Institutional Review Boards – Topic Section Discussions*, National Institutes of Health Bioethics Interest Group, Rockville, Maryland. Available at: http://www.nih.gov/sigs/bioethics/reports/topics.htm [Accessed 04/09/02 – No longer available].
Eyler, L. T. and Jeste, D. V. (2006). 'Enhancing the Informed Consent Process: A Conceptual Overview'. *Behavioral Sciences and the Law*, vol. 24(4), pp. 553–568.
Faden, R. R., Beauchamp, T. L. and King, N. M. P. (1986). *A History and Theory of Informed Consent*, Oxford University Press, New York.
Farah, M. J. (2002). 'Emerging ethical issues in neuroscience'. *Nature Neuroscience*, vol. 5(2), pp. 1123–1129.
Farah, M. J., Illes J., Cook-Deegan, R., Gardner, H., Kande, I. E., King, P., Parens, E., Sahakian, B. and Wolpe, P. R. (2004). 'Neurocognitive Enhancement: What Can We Do and What Should We Do?' *Nature Reviews Neuroscience*, vol. 5, pp. 421–425.
Farah, M. J., Wolpe, P. R. and Caplan, A. (2005). 'Brain Research and Neuroethics'. In: J. Gunning and S. Holm (eds.) *Ethics, Law and Society: Volume: 1*, Aldershot, Ashgate, pp. 261–264.
Faro, S. H. and Mohamed, F. B. (eds.) (2006). *Functional MRI: Basic Principles and Clinical Applications*, Springer, Heidelberg.
Feenan, D. (1997). 'Capable People: Empowering the Patient in the Assessment of Capacity'. *Health Care Analysis*, vol. 5(3), pp. 227–236.
Feinberg, J. (1980). *Rights, Justice, and the Bounds of Liberty: Essays in Social Philosophy*, Princeton University Press, Princeton, NJ.
Feinberg, J. (1986). *The Moral Limits of the Criminal Law: Harm to Self*, Oxford University Press, New York.

Fischman, M. W. (2000). 'Informed Consent'. In: B. Sales and S. Folkman (eds.) *Ethics in Research with Human Participants*, American Psychiatric Association, Washington, DC, pp. 35–48.
Fletcher, J. C. (1983). 'The Evolution of the Ethics of Informed Consent'. In: K. Berg and K. E. Tranøy (eds.) *Research Ethics* (Progress in Clinical and Biological Research Volume 128), Alan R. Liss, New York, pp. 187–228.
Flory, J. and Emanuel, E. (2004). 'Interventions to Improve Research Participants' Understanding in Informed Consent for Research: A Systematic Review'. *Journal of the American Medical Association*, vol. 292(13), pp. 1593–1601.
Fontana, A. and Smith, R. W. (1989). 'Alzheimer's Disease Victims: The 'Unbecoming' of Self and the Normalization of Competence'. *Sociological Perspectives*, vol. 32(1), pp. 35–46.
Ford, M. (2005). 'The Personhood Paradox and the Right to Die'. *Medical Law Review*, vol. 13, pp. 80–101.
Foster, C. (2001). *The Ethics of Medical Research on Humans*, Cambridge University Press, Cambridge.
Fox, R. C. (1959). *Experiment Perilous*, Free Press, Glencoe, IL.
Frankfurt, H. G. (1971). 'Freedom of the Will and the Concept of a Person'. *Journal of Philosophy*, vol. 68, pp. 5–20.
Freedman, B. (1981). 'Competence, Marginal and Otherwise: Concepts and Ethics'. *International Journal of Law and Psychiatry*, vol. 4, pp. 53–72.
Freund, P. A. (1972). 'Legal Frameworks for Human Experimentation'. In: P. A. Freund (ed.) *Experimentation with Human Subjects*, George Allen & Unwin, London, pp. 105–115.
Frith, U. (2004). 'Autism and Asperger's Syndrome'. In: R. L. Gregory (ed.) *The Oxford Companion to the Mind*, second edition, Oxford University Press, Oxford, pp. 79–80.
Fulford, K. W. M. and Howse, K. (1993). 'Ethics of Research with Psychiatric Patients: Principles, Problems and the Primary Responsibilities of Researchers'. *Journal of Medical Ethics*, vol. 19, pp. 85–91.
Fundudis, T. (2003). 'Consent Issues in Medico-Legal Procedures: How Competent are Children to Make Their Own Decisions?' *Child and Adolescent Mental Health*, vol. 8(1), pp. 18–22.
Garrett, R. (1994). 'The Problem of Despair'. In: G. Graham and G. L. Stephens (eds.) *Philosophical Psychopathology*, MIT, Cambridge, MA, pp. 73–89.
Gatz, M. (2006). 'Cognitive Capacities of Older Adults Who Are Asked to Consent to Medical Treatment or to Clinical Research'. *Behavioral Sciences and the Law*, vol. 24(4), pp. 465–468.
Gauthier, D. (1986). *Morals by Agreement*, Clarendon, Oxford.
Gauthier. D. (1988). 'Morality, Rational Choice and Semantic Representation: A Reply to My Critics'. In: E. F. Paul, F. D. Miller Jr. and J. Paul (eds.) *The New Social Contract: Essays on Gauthier*, Blackwell, Oxford: pp. 173–189.
Gauthier, D. (1991). 'Rational Constraint: Some Last Words'. In: P. Vallentyne (ed.) *Contractarianism and Rational Choice: Essays on David Gauthier's Morals By Agreement*, Cambridge University Press, New York, pp. 323–330.
General Medical Council (2001). *Medical Research: The Role and Responsibilities of Doctors* (Draft version), General Medical Council, London.
General Medical Council (2002). *Research: The Role and Responsibilities of Doctors,* General Medical Council, London. Available at: http://www.gmc-uk.org/guidance/current/library/research.asp [Accessed 30/03/08].
George, M. S. and Belmaker, R. H. (2000). *Transcranial Magnetic Stimulation in Neuropsychiatry*, American Psychiatric Press, Washington, DC.
Geula, C. (1998). 'Abnormalities of neural circuits in Alzheimer's disease: hippocampus and cortical cholinergic innervation'. *Neurology*, vol. 51, pp. 18–29.
Gewirth, A. (1960). 'Positive 'Ethics' and Normative 'Science'. *Philosophical Review,* vol. 69, pp. 311–330.
Gewirth, A. (1978). *Reason and Morality*, The University of Chicago Press, Chicago, IL.
Gewirth, A. (1981). 'Are There Any Absolute Rights?' *Philosophical Quarterly*, vol. 31, pp. 1–16.

Gewirth, A. (1982a). 'The Basis and Content of Human Rights'. In: A. Gewirth (ed.) *Human Rights: Essays on Justification and Applications*, The University of Chicago Press, Chicago, pp. 41–78.

Gewirth, A. (1982b). 'The Is-Ought Problem Resolved'. In: A. Gewirth (ed.) *Human Rights: Essays on Justification and Applications*, The University of Chicago Press, Chicago, IL, pp. 100–127.

Gewirth, A. (1984a). 'The Epistemology of Human Rights'. *Social Philosophy and Policy*, vol. 1 (2), pp. 1–24.

Gewirth, A. (1984b). 'Replies to My Critics'. In: E. Regis Jr. (ed.) *Gewirth's Ethical Rationalism: Critical Essays with a Reply by Alan Gewirth*, The University of Chicago Press, Chicago, IL, pp. 192–255.

Gewirth, A. (1996). *The Community of Rights*, The University of Chicago Press, Chicago, IL.

Gewirth, A. (1998). *Self-Fulfilment*. Princeton University Press, Princeton, NJ.

Giasson, B. I., Forman, M. S., Higuchi, M., Golbe, L. I., Graves, C. L., Kotzbauer, P. T., Trojanowski, J. Q. and Lee, V. M. Y. (2003). 'Initiation and Synergistic Fibrillization of Tau and Alpha-Synuclein'. *Science*, vol. 300, no. 5619, pp. 636–640.

Gillberg, C. (2002). *A Guide to Asperger's Syndrome*, Cambridge University Press, Cambridge.

Gillick v West Norfolk and Wisbech Area Health Authority [1986] A.C. 112.

Gilligan, C. (1982). *In a Different Voice: Psychological Theory and Women's Development*, Harvard University Press, Cambridge, MA.

Glannon, W. (2007). *Bioethics and the Brain*, Oxford University Press, New York.

Glantz, L. H. (1994). 'The Law of Human Experimentation with Children'. In: M. A. Grodin and L. A. Glantz (eds.). *Children as Research Subjects: Science, Ethics and Law*, Oxford University Press, New York, pp. 103–130.

Glantz, L. H. (1998). 'Research with Children'. *American Journal of Law and Medicine*, vol. 24(2/3), pp. 213–244.

Glass, K. C. (1997). 'Refining Definitions and Devising Instruments: Two Decades of Assessing Mental Competence'. *International Journal of Law and Psychiatry*, vol. 20(1), pp. 5–33.

Glover, J. (1988). *I: The Philosophy and Psychology of Personal Identity*, Penguin, Harmondsworth.

Goffman, E. (1961). *Asylums: Essays on the Social Situation of Mental Patients and Other Inmates*, Penguin, Harmondsworth.

Goldstein, W. M. and Hogarth, R. M. (1997). *Research on Judgment and Decision Making: Currents Connections and Controversies*, Cambridge University Press, Cambridge.

Goodin, R. E. (1985). *Protecting the Vulnerable: A Re-analysis of Our Social Responsibilities*, The University of Chicago Press, Chicago, IL.

Goodpaster, K. E. (1978). 'On Being Morally Considerable'. *The Journal of Philosophy*, vol. 75(6), pp. 308–325.

Goodwin, F. K. and Ghaemi, S. N. (2000). 'An Introduction to and Historical Review of Mood Disorders'. In: M. G. Gelder, J. J. López-Ibor Jr. and N. C. Andreasen (eds.) *New Oxford Textbook of Psychiatry: Volume 1*, Oxford University Press, Oxford, pp. 677–682.

Green, M. D. (1941). 'Judicial Tests of Mental Competency'. *Missouri Law Review*, vol. 6, pp. 141–165.

Grinnell, F. (2004). 'Subject Vulnerability: The Precautionary Principle of Human Research'. *The American Journal of Bioethics*, vol. 4(3), pp. 72–74.

Grisso, T. (ed.) (2002). *Evaluating Competencies: Forensic Assessments and Instruments*, Springer, Berlin.

Grisso, T. and Appelbaum, P. S. (1995). 'The MacArthur Treatment Competence Study, III: Abilities of Patients to Consent to Psychiatric and Medical Treatments'. *Law and Human Behavior*; vol. 19, pp. 149–174.

Grisso, T. and Appelbaum, P. S. (1998). *Assessing Competence to Consent to Treatment: A Guide for Physicians and Other Health Professionals*, Oxford University Press, New York.

Grisso, T, Appelbaum P. S., Mulvey, E. and Fletcher, K. (1995). 'The MacArthur Treatment Competence Study II: Measure of Abilities Related to Competence to Consent to Treatment'. *Law and Human Behavior*, vol. 19, pp. 127–148.

Guinn, D. E. (2002). 'Mental Competence, Caregivers and the Process of Consent: Research Involving Alzheimer's Patients and Others with Decreasing Mental Capacity'. *Cambridge Quarterly of Healthcare Ethics*, vol. 11, pp. 230–245.

Gunn, M. J., Wong, J. G., Clare, I. C. H. and Holland, A. J. (1999). 'Decision-Making Capacity'. *Medical Law Review*, vol. 7, pp. 269–306.

Häfner, H. and Wolpert, E. M. (1996). *New Research in Psychiatry*, Hogrefe & Huber, Seattle, WA.

Hall, L. L. (1996). *Genetics and Mental Illness: Evolving Issues for Research and Society*, Plenum, New York.

Hanin, I. Cacabelos, R. and Fisher, A. (2005). *Recent Progress in Alzheimer's and Parkinson's Diseases*, Taylor & Francis, London.

Harré, R. (1984). *Personal Being*, Harvard University Press, Cambridge, MA.

Harris, G. W. (1997). *Dignity and Vulnerability: Strength and Quality of Character*, University of California Press, Berkeley, CA.

Harris, J. (1985). *The Value of Life: An Introduction to Medical Ethics*, Routledge, London.

Harris J. (1999). 'The Concept of the Person and the Value of Life'. *Kennedy Institute of Ethics Journal,* vol. 9(4), pp. 293–308.

Harris, J. (2005a). 'Scientific research is a moral duty'. *Journal of Medical Ethics*, vol. 31, pp. 242–248.

Harris, J. (2005b). 'The Right to Die Lives! There Is No Personhood Paradox'. *Medical Law Review*, vol. 13, pp. 386–392.

Harris J. and Holm S. (2002). 'Extending Human Lifespan and the Precautionary Paradox'. *Journal of Medicine and Philosophy*, vol. 27(3), pp. 355–368.

Harris, J. and Holm, S. (2003). 'Should We Presume Moral Turpitude in Our Children? – Small Children and Consent to Medical Research'. *Theoretical Medicine and Bioethics*, vol. 24, pp. 121–129.

Harris, P. A. (2002). *The Distinction Between Law and Ethics in Natural Law Theory*, Edwin Mellen, Lewiston, NY.

Haksar, V. (1991). *Indivisible Selves and Moral Practice*, Edinburgh University Press, Edinburgh.

Häyry, M. (2005). 'Precaution and Solidarity'. *Cambridge Quarterly of Healthcare Ethics*, vol. 14, pp. 199–206.

Heinimaa, M. (2000). 'Ambiguities in the Psychiatric Use of the Concepts of the Person: An Analysis'. *Philosophy, Psychiatry and Psychology*, vol. 7(2), pp. 125–136.

Hill, E. L. and Frith, U. (2004). 'Understanding Autism: Insights from Mind and Brain'. In: U. Frith and E. L. Hill (eds.) *Autism: Mind and Brain*, Oxford University Press, Oxford, pp. 1–20.

Hill, J. F. (1984). 'Are Marginal Agents Our Recipients?' In: E. Regis Jr. (ed.) *Gewirth's Ethical Rationalism: Critical Essays with a Reply by Alan Gewirth*, The University of Chicago Press, Chicago, IL, pp. 180–191.

Hinterberger, T., Birbaumer, N. and Flor, H. (2005). 'Assessment of Cognitive Function and Communication Ability in a Completely Locked-in Patient'. *Neurology*, vol. 64(7), pp. 1307–1308.

Hoffman R. E., Boutros N. N., Hu, S., Berman, R. M., Krystal, J. H. and Charney, D. S. (2000). 'Transcranial Magnetic Stimulation and Auditory Hallucinations in Schizophrenia'. *Lancet*, vol. 355, pp. 1073–1075.

Hoffmaster, B. (2006). 'What Does Vulnerability Mean?' *Hastings Center Report*, vol. 36(2), pp. 38–45.

Hohfeld, W. N. (2001) [1919]. *Fundamental Legal Conceptions as Applied in Judicial Reasoning*, D. Campbell, P. Thomas and N. E. Simmonds (eds.), Dartmouth, London.

Holm, S. (2001). 'Autonomy, Authenticity or Best Interest: Everyday Decision-Making and Persons with Dementia'. *Medicine, Health Care and Philosophy,* vol. 4, pp. 153–159.

Holtzheimer P. E. and Nemeroff C. B. (2006). 'Advances in the Treatment of Depression'. *NeuroRx: The Journal of the American Society for Experimental NeuroTherapeutics*, vol. 3(1), pp. 42–56.

Hougham, G. W., Sachs, G. A., Danner, D., Mintz, J., Patterson, M., Roberts, L. W., Siminoff, L. A., Sugarman, J., Whitehouse, P. J. and Wirshing. D. (2003). 'Empirical Research on Informed Consent with the Cognitively Impaired'. *IRB: Ethics and Human Research,* vol. 25(5), pp. 26–32.

House of Lords and House of Commons Joint Committee on the Draft Mental Incapacity Bill (2003). *Draft Mental Incapacity Bill, Volume I - Report Together with Formal Minutes*. TSO, London. Available at: http://www.publications.parliament.uk/pa/jt200203/jtselect/jtdmi/189/189.pdf [Accessed 30/03/08].

Howe, V., Foster, K., Jenkins, K., Keks, N., Skene, L. and Copolov, D. (2003). 'Competence to Give Informed Consent to Research Participation in Persons with Schizophrenia and Related Psychosis'. *Psychiatry, Psychology and Law*, vol. 10(1), pp. 85–90.

Hughes, J. C. (2001). 'Views of the Person with Dementia'. *Journal of Medical Ethics*, vol. 27, pp. 86–91.

Hughes, J. C., Louw, S. J. and Sabat, S. R. (eds.) (2006). *Dementia: Mind, Meaning and the Person*, Oxford University Press, Oxford.

Human Tissue Act 2004 (c. 30). TSO, London.

Hurst, S. A. (2008). 'Vulnerability in Research and Health Care: Describing the Elephant in the Room?' *Bioethics*, vol. 22(4), pp. 191–202.

Huthwaite, J. S., Martin, R. C., Griffith, H. R., Anderson, B., Harrell, L. E. and Marson, D. C. (2006). 'Declining Medical Decision-Making Capacity in Mild AD: A Two-Year Longitudinal Study'. *Behavioral Sciences and the Law*, vol. 24(4), pp. 453–463.

Illes, J. (ed.) (2006). *Neuroethics: Defining the Issues in Theory, Practice and Policy*, Oxford University Press, New York.

In Re Conroy 486 A.2d 1209 (N.J. 1985).

In Re Maida Yetter (1973) 62 Pa. D. & C.2d 619 (Pa.Com.Pl.)

In Re Quackenbush (1978) 156 NJ Super 282, 383 A 2d 785 (Morris County Ct).

In Re Roe 583 N.E.2d 1282, 1286 (Mass. 1992).

In Re Schiller 148 N.J.Super. 168, 372 A.2d 360 (N.J.Super.Ch. Feb 4, 1977).

In Re Virgil D 524 N.W.2d 894, 895 (Wis. 1994).

International Conference on Harmonisation of Technical Requirements for Registration of Pharmaceuticals for Human Use (ICH) (1996). *ICH Harmonised Tripartite Guideline for Good Clinical Practice*. Geneva, Switzerland. Available at: http://www.ich.org/LOB/media/MEDIA482.pdf [Accessed 30/03/08].

Iris (directed by Richard Eyre). BBC Films, 2001.

Issa, A. M. and Keyserlingk, E. W. (2000). 'Current and Future Clinical Trials for Alzheimer's Disease: Evolving Ethical Concerns'. *Progress in Neuro-Psychopharmacology and Biological Psychiatry*, vol. 24, pp. 1229–1249.

Jablensky, A. (2000). 'Course and Outcome of Schizophrenia and Their Prediction'. In: M. G. Gelder, J. J. López-Ibor Jr. and N. C. Andreasen (eds.) *New Oxford Textbook of Psychiatry: Volume 1*, Oxford University Press, Oxford, pp. 612–621.

Jackson, E. and Warner, J. (2002). 'How Much Do Doctors Know About Consent and Capacity?' *Journal of the Royal Society of Medicine*, vol. 95, pp. 601–603.

Jansen, S. (2000). 'Recommendation No. R (99) 4 of the Committee of Ministers to Member States on Principles Concerning the Legal Protection of Incapable Adults, an Introduction in Particular to Part V Interventions in the Health Field'. *European Journal of Health Law*, vol. 7, pp. 333–347.

Jaspers, K. (1997a) [1913]. *General Psychopathology, Volume I* (trans. J. Hoenig and M. W. Hamilton). The John Hopkins University Press, Baltimore, MD.

Jaspers, K. (1997b) [1913]. *General Psychopathology, Volume II* (trans. J. Hoenig and M. W. Hamilton). The John Hopkins University Press, Baltimore, MD.

Jaworska, A. (1999). 'Respecting the Margins of Agency: Alzheimer's Patients and the Capacity to Value'. *Philosophy and Public Affairs*, vol. 28(2), pp. 105–138.

Jaworska, A. (2006). 'Ethical Issues in Neurodegenerative Disease: Respecting Patients at the Twilight of Agency'. In: J. Illes (ed.) *Neuroethics: Defining the Issues in Theory, Practice and Policy*, Oxford University Press, New York, pp. 87–101.

Jellinger, K. A., Schmidt, R. and Widisch, M. (eds.) (2000). *Advances in Dementia Research*, Springer, Vienna.

Jeste, D. V. Dunn, L B., Palmer, B. W. Saks, E. Halpain, M. Cook, A. Appelbaum P. S., Schneiderman L. (2003). 'A Collaborative Model for Research on Decisional Capacity and Informed Consent in Older Patients with Schizophrenia: Bioethics Unit of a Geriatric Psychiatry Intervention Research Center'. *Psychopharmacology*, vol. 171, pp. 68–74.

Joffe, S., Francis-Cook, E., Cleary, P. D., Clark, J. W., Weeks, J. C. (2001). 'Quality of Informed Consent: A New Measure of Understanding Among Research Subjects'. *Journal of the National Cancer Institute*, vol. 3 (2): 139–147.

Jonas, M. F. (2007). 'Competence to Consent'. In: R. E. Ashcroft, A. Dawson, H. Draper and J. R. McMillan (eds.) *Principles of Health Care Ethics*, second edition, Wiley, Chichester, pp. 255–262.

Jonsen, A. R. (1998). *The Birth of Bioethics*, Oxford University Press, New York.

Kahn, J. P., Mastroianni, A. C. and Sugarman, J. (eds.) (1998). *Beyond Consent: Seeking Justice in Research*, Oxford University Press, New York.

Kaimowitz v Department of Mental Health of the State of Michigan (1973). 42 USLW 2063 (Mich Cir Ct).

Kant, I. (1991) [1785]. *The Moral Law: Groundwork of the Metaphysic of Morals* (trans. and analysed by H. J. Paton). Routledge, London.

Karlawish, J. (2007). 'Research On Cognitively Impaired Adults'. In: B. Steinbock (ed.) *The Oxford Handbook of Bioethics*, Oxford University Press, Oxford, pp. 597–620.

Katz, J. (1972). *Experimentation with Human Beings*, Russell Sage Foundation, New York.

Katz, J. (1993). 'Human Experimentation and Human Rights'. *Saint Louis University Law Journal*, vol. 38, pp. 7–54.

Katz, J. (2002). *The Silent World of Doctor and Patient* (reissued with a foreword by A. M. Capron), The Johns Hopkins University Press, Baltimore, MD.

Keenan, T. (2001). *An Introduction to Child Development*, Sage, Thousand Oaks, CA.

Keith-Spiegel, P. (1983). 'Children and Consent to Participation in Research'. In: Melton, G. B., Koocher, G. P. and Saks, M. J. (eds.) *Children's Competence to Consent*, Plenum, New York, pp. 179–211.

Kemp, P. (ed.) (1999). *Basic Ethical Principles in Bioethics and Biolaw, 1995–1998 Part B*. Council of Europe, Strasbourg. Available at: http://ec.europa.eu/research/biosociety/pdf/final_rep_95_0207.pdf [Accessed 30/03/08].

Kennedy, I. and Grubb, A. (2000). *Medical Law*, third edition, Butterworths, London.

Kim, S. Y. H., Caine, E. D., Currier, G. W. M., Leibovici, A. and Ryan, J. M. (2001). 'Assessing the Competence of Persons with Alzheimer's Disease in Providing Informed Consent for Participation in Research'. *American Journal of Psychiatry*, vol. 158(5), pp. 712–717.

Kim, S. Y. H., Karlawish, J. H. T., Caine, E. D. (2002a). 'Current State of Research on Decision-Making Competence of Cognitively Impaired Elderly Persons'. *American Journal of Geriatric Psychiatry*, vol. 10(2), pp. 151–165.

Kim, S. Y. H., Cox, C., Caine, E. D. (2002b). 'Impaired Decision-Making Ability in Subjects with Alzheimer's Disease and Willingness to Participate in Research'. *American Journal of Psychiatry*, vol. 159, pp. 797–802.

King, N. M. P., Henderson, G. E., Stein, J. (eds.) (1999). *Beyond Regulations: Ethics in Human Subjects Research*, The University of North Carolina Press, Chapel Hill, NC.

Kipnis, K. (2003). 'Seven Vulnerabilities in the Pediatric Research Subject'. *Theoretical Medicine and Bioethics*, vol. 24, pp. 107–120.

Kipnis, K. (2004). 'Vulnerability in Research Subjects: An Analytical Approach'. In: D. C. Thomasma and D. N. Weisstub (eds.) *The Variables of Moral Capacity*, Kluwer Academic, Dordrecht, pp. 217–231.

Kitwood, T. (1997). *Dementia Reconsidered: The Person Comes First*, Open University Press, Buckingham.

Kohlberg, L. (1984). *The Psychology of Moral Development: The Nature and Validity of Moral Stages*, Harper & Row, San Francisco, CA.

Kopelman, L. M. (1990). 'On the Evaluative Nature of Competence and Capacity Judgments'. *International Journal of Law and Psychiatry*, vol. 13, pp. 309–329.
Kopelman, L. M. (2004). 'Adolescents as Doubly Vulnerable Research Subjects'. *American Journal of Bioethics*, vol. 4(1), pp. 50–52.
Kopelman, L. M., Resnik, D. and Weed, D. L. (2004). 'What is the Role of the Precautionary Principle in the Philosophy of Medicine and Bioethics?' *Journal of Medicine and Philosophy*, vol. 29(3), pp. 255–258.
Koren, G., Birenbaum-Carmeli, D., Carmeli, Y. S. and Haslam, R. (1993). 'Maturity of Children to Consent to Medical Research: The Babysitter Test'. *Journal of Medical Ethics*, vol. 19, pp. 142–147.
Korsgaard, C. M. (1989). 'Personal Identity and the Unity of Agency: A Kantian Response to Parfit'. *Philosophy and Public Affairs*, vol. 18(2), pp. 101–132.
Kottow, M. H. (2003). 'The Vulnerable and the Susceptible'. *Bioethics*, vol. 17(5–6), 460–471.
Kottow, M. H. (2004). 'Vulnerability: What kind of principle is it?' *Medicine, Health Care and Philosophy*, vol. 7, pp. 281–287.
Kramer, M. H., Simmonds, N. E. and Steiner, H. (1998). *A Debate Over Rights: Philosophical Enquiries*, Clarendon, Oxford.
Krimsky, S. (2003). *Science in the Private Interest: Has the Lure of Profits Corrupted Biomedical Research?* Rowman & Littlefield, Maryland.
Lacan, J. (2006). *Écrits* (trans. B. Fink). W. W. Norton, New York.
Laing, R. D. (1971). *The Politics of the Family and Other Essays*, Tavistock, London.
Laing, R. D. (1990) [1960]. *The Divided Self: An Existential Study in Sanity and Madness*, Penguin, Harmondsworth.
Landry, L. Y. (1999). 'Multi-Disciplinary Competence Assessment: A Case Study in Consensus and Culture'. *Theoretical Medicine and Bioethics*, vol. 20, pp. 423–499.
Lane v Candura 376 N.E.2d 1232 (Mass. App. Ct. 1978).
Lapid, M. I., Rummans, T. A., Pankratz, V. S., Appelbaum, P. S. (2004). 'Decisional Capacity of Depressed Elderly to Consent to Electroconvulsive Therapy'. *Journal of Geriatric Psychiatry and Neurology*, vol. 17(1), pp. 42–46.
Lavery, A. (03/05/03) *'Alzheimer's Vitamins Trial Begins'*. BBC News Online. Available at: http://news.bbc.co.uk/1/hi/health/2996807.stm. [Accessed 30/03/08].
Law Commission of England and Wales (1991). *Mental Incapacitated Adults and Decision-Making: An Overview*, Consultation Paper 119, HMSO, London.
Law Commission of England and Wales (1993). *Mentally Incapacitated Adults and Decision-Making: Medical Treatment and Research*, Consultation Paper 129, HMSO, London.
Law Commission of England and Wales (1995). *Mental Incapacity*, Report 231, HMSO, London.
Lederer, S. E. (1997). *Subjected to Science: Human Experimentation in America Before the Second World War*, The Johns Hopkins University Press, Baltimore, MD.
Leff, J. P. and Wing, J. K. (1971). 'Trial of Maintenance Therapy in Schizophrenia'. *British Medical Journal*, vol. 775(3), pp. 599–604.
Levinas, E. (1981). *Otherwise than Being: Or Beyond Essence* (trans. A. Lingis). Kluwer, Dordrecht.
Levine, C., Faden, R., Grady, C., Hammerschmidt, D., Eckenwiler, L. and Sugarman, J. (2004). 'The Limitations of "Vulnerability" as a Protection for Human Research Participants'. *The American Journal of Bioethics*, vol. 4(3), pp. 44–49.
Levine, R. J. (1988). *Ethics and Regulation of Clinical Research*, second edition, Yale University Press, New Haven, CT.
Lewis, C. (1981). 'How Adolescents Approach Decisions: Changes over Grades Seven to Twelve and Policy Implications'. *Child Development*, vol. 52, pp. 538–544.
Lifton, R. J. (1988). *The Nazi Doctors: Medical Killing and the Psychology of Genocide*, Basic Books, New York.
Lindley, R. (1986). *Autonomy*, Macmillan, London.

Lo, B. (1990). 'Assessing Decision-Making Capacity'. *Law, Medicine and Health Care*, vol. 18: 192–201.
Locke, J. (1997) [1689]. *An Essay Concerning Human Understanding* (ed. Roger Woolhouse), Penguin, Harmondsworth.
Lomasky, L. E. (1987). *Persons, Rights, and the Moral Community*, Oxford, New York.
Lord Chancellor's Department (1997). *Who Decides? Making Decisions on Behalf of Mentally Incapacitated Adults*. Cm. 3808, HMSO, London.
Lord Chancellor's Department (1999). *Making Decisions: Government Proposals for Making Decisions on Behalf of Mentally Incapacitated Adults*. Cm. 4465, HMSO, London.
Lovestone, S. (2000). 'Dementia: Alzheimer's Disease'. In: M. G. Gelder, J. J. López-Ibor Jr and N. C. Andreasen (eds.) *New Oxford Textbook of Psychiatry: Volume 1*, Oxford University Press, Oxford, pp. 387–397.
Lyons, D. (1993). 'On Formal Justice'. In: D. Lyons (ed.) *Moral Aspects of Legal Theory: Essays on Law, Justice and Political Responsibility*, Cambridge University Press, Cambridge, pp. 13–40.
MacIntyre, A. (1999). *Dependent Rational Animals: Why Human Beings Need the Virtues*. Open Court, Chicago, IL.
Macklin, R. (1983). 'Philosophical Conceptions of Rationality and Psychiatric Notions of Competency'. *Synthese*, vol. 57, pp. 205–224.
Macklin, R. (2004). *Double Standards in Medical Research in Developing Countries*, Cambridge University Press, Cambridge.
Maio, G. (2002). 'The Relevance of Rawls's Principle of Justice for Research on Cognitively Impaired Patients'. *Theoretical Medicine and Bioethics*, vol. 23, pp. 45–53.
Malcolm, N. (1958). 'Knowledge of Other Minds'. *The Journal of Philosophy*, vol. 55(23), pp. 969–978.
Malhi, G. S. and Sachdev, P. (2002). 'Novel Physical Treatments for The Management of Neuropsychiatric Disorders'. *Journal of Psychosomatic Research*, vol. 53, pp. 709–719.
Marcus, S. J. (ed.) (2002). *Neuroethics: Mapping the Field*, Dana, New York.
Marson, D. C. (2001). 'Loss of Competency in Alzheimer's Disease: Conceptual and Psychometric Approaches'. *International Journal of Law and Psychiatry*, vol. 24, pp. 267–283.
Marson, D. C., Hawkins L, McInturff, B, Harrell L. E. (1997). 'Cognitive Models that Predict Physician Judgments of Capacity to Consent in Mild Alzheimer's Disease'. *Journal of the American Geriatric Society*, vol. 45(4), pp. 458–464.
Marson, D. C., Schmitt, F., Ingram, K. K., Harrell, L. E. (1994). 'Determining the Competency of Alzheimer's Patients to Consent to Treatment and Research'. *Alzheimer's Disease and Associated Disorders*, vol. 8 (supplement 4), pp. 5–18.
Mason, J. K. and Laurie, G. T. (2006). *Mason and McCall Smith's Law and Medical Ethics*, seventh edition, Oxford University Press, Oxford.
McCabe, M. A. (1996). 'Involving Children and Adolescents in Medical Decision Making: Developmental and Clinical Considerations'. *Journal of Paediatric Psychology*, vol. 21(4), pp. 505–516.
McCance, R. A. (1951). 'The Practice of Experimental Medicine'. Reproduced in S. J. Reiser, A. J. Dyck and W. J. Curran (eds.) (1977) *Ethics in Medicine: Historical Perspectives and Contemporary Concerns*, MIT, Cambridge, MA, pp. 274–277.
McCarron, M. C. (1990). 'The Right to Refuse Antipsychotic Drugs: Safeguarding the Mentally Incompetent Patient's Right to Procedural Due Process'. *Marquette Law Review*, vol. 73, pp. 477–512.
McConnell, T. (2000). *Inalienable Rights: The Limits of Consent in Medicine and the Law*, Oxford University Press, New York.
McMillan, J. (2006). 'Identity: Self and Dementia'. In: J. C. Hughes, S. J. Louw and S. R. Sabat (eds.) *Dementia: Mind, Meaning and the Person*, Oxford University Press, Oxford, pp. 63–70.
McMillan, J. and Gillett, G. (2002). 'Consent as Empowerment: The Roles of Postmodern and Narrative Ethics'. In: K. W. M. Fulford, D. L. Dickenson and T. H. Murray (eds.) *Health Care Ethics and Human Values*, Blackwell, Oxford, pp. 225–228.

McNeill, P. M. (1993). *The Ethics and Politics of Human Experimentation*, Cambridge University Press, Cambridge.

Mead, S., Joiner, S., Desbruslais, M., Beck, J. A., O'Donoghue, M., Lantos, P., Wadsworth, J. D. F. and Collinge, J. (2007). 'Creutzfeldt-Jakob Disease, Prion Protein Gene Codon 129VV, and a Novel PrP^{Sc} Type in a Young British Woman'. *Archives of Neurology*, vol. 64(12), pp. 1780–1784.

Medical Research Council Working Party on the Mentally Incapacitated (1991). *The Ethical Conduct of Research on the Mentally Incapacitated*, Medical Research Council, London.

Medical Research Council (2004). *Medical Research Involving Children*. Medical Research Council, London. Available at: http://www.mrc.ac.uk/Utilities/Documentrecord/index.htm?d = MRC002430 [Accessed 30/03/08].

Medical Research Council (2007). *MRC Ethics Guide 2007: Medical Research Involving Adults Who Cannot Consent*. Medical Research Council, London. Available at: http://www.mrc.ac.uk/Utilities/Documentrecord/index.htm?d = MRC002409 [Accessed 30/03/08].

Meikle, J. (24/12/02). 'CJD Sufferer Cleared to Have Pioneering Surgery'. *The Guardian* [UK]. Available at: http://www.guardian.co.uk/print/0,3858,4572986–103527,00.html [Accessed 30/03/08].

Meisel, A. and Kuczewski, M. (1996). 'Legal and Ethical Myths about Informed Consent'. *Archives of Internal Medicine*, vol. 156, pp. 2521–2526.

Mental Capacity Act 2005 (c. 9). TSO, London.

Mental Health Act 2007 (c. 12). TSO, London.

Merkel, R., Boer, G., Fegert, J., Galert, T., Hartmann, D., Nuttin, B. and Rosahl, S. (2007). *Intervening in the Brain: Changing Psyche and Society*, Springer, Berlin.

Meslin, E. M. (2001). 'International Regulation, Informed Consent and Medical Research: A Perspective from the USA and Canada'. In: L. Doyal and J. S. Tobias (eds.) *Informed Consent in Medical Research*, BMJ Books, London, pp. 148–155.

Mill, J. S. (1991) [1859]. *On Liberty and Other Essays* (edited with an introduction and notes by J. Gray). Oxford University Press, Oxford.

Miller, V. A., Drotar, D. and Kodish, E. (2004). 'Children's Competence for Assent and Consent: A Review of Empirical Findings'. *Ethics and Behavior*, vol. 14(3), pp. 255–295.

Misra, S. and Ganzini, L. (2004). 'Capacity to consent to research among patients with bipolar disorder'. *Journal of Affective Disorders*, vol. 80, pp. 115–123.

Mitchell, P. (1997). *Introduction to Theory of Mind: Children, Autism and Apes*. Arnold, London.

Mitscherlich, A. and Mielke, F. (1962). *Medizin ohne Menschlichkeit* (translated as: *The Death Doctors*), Elek Books, London.

Moreno, J. D. (2001). *Undue Risk: Secret State Experiments on Humans*, Routledge, London.

Morgenbesser, S. (1977). 'Experimentation and Consent: A Note'. In: S. F. Spicker and H. T. Engelhardt Jr. (eds.) *Philosophical Medical Ethics: Its Nature and Significance*. D. Reidel, Dordrecht, pp. 96–110.

Morreim, E. H. (1983). 'Three Concepts of Patient Competence'. *Theoretical Medicine*, vol. 4, pp. 231–51.

Morreim, E. H. (1991). 'Competence: At the Intersection of Law, Medicine and Philosophy'. In: M. A. G. Cutter and E. E. Shelp (eds.) *Competency: A Study of Informal Competency Determinations in Primary Care*, Kluwer Academic, Dordrecht, pp. 93–125.

Morris, C. D., Niederbuhl, J. M. and Mahr, J. M. (1993). 'Determining the Capability of Individuals with Mental Retardation to Give Informed Consent'. *American Journal on Mental Retardation*, vol. 98, pp. 263–272.

Morrison, L. J. (2005). *Talking Back to Psychiatry: The Psychiatric Consumer/Survivor/Ex-Patient Movement*, Routledge, New York.

Moser, D. J., Schultz, S. K., Arndt, S., Benjamin, M. L., Fleming, F. W., Brems, C. S., Paulsen, J. S., Appelbaum, P. S. and Andreasen, N. C. (2002). 'Capacity to Provide Informed Consent for Participation in Schizophrenia and HIV Research'. *American Journal of Psychiatry*, vol. 159, pp. 1201–1207.

Moser, D. J., Reese, R. L., Hey, C. T., Schultz, S. K., Arndt, S., Beglinger, L. J., Duff, K. M. and Andreasen, N. C. (2006). 'Using a Brief Intervention to Improve Decisional Capacity in Schizophrenia Research'. *Schizophrenia Bulletin,* vol. 32(1), pp. 116–120.

Mulvey, E. and Peeples, F. (1996). 'Are Disturbed and Normal Adolescents Equally Competent to Make Decisions About Mental Health Treatments?' *Law and Human Behavior*, vol. 20, pp. 273–287.

Murphy, J. G. (1979a). 'Incompetence and Paternalism'. In: J. G. Murphy (ed.) *Retribution, Justice and Therapy: Essays in the Philosophy of Law*, D. Reidel, Dordrecht, pp. 165–182.

Murphy, J. G. (1979b). 'Therapy and the Problem of Autonomous Consent'. *International Journal of Law and Psychiatry*, vol. 2(4), pp. 415–430.

Musalek, M. (2003). 'Meaning and Causes of Delusions'. In: B. Fulford, K. Morris, J. Sadler and G. Stanghellini (eds.) *Nature and Narrative: An Introduction to the New Philosophy of Psychiatry*, Oxford University Press, Oxford, pp. 155–169.

National Commission for the Protection of Human Subjects of Biomedical and Behavioral Research (1979). *The Belmont Report: Ethical Principles and Guidelines for the Protection of Human Subjects of Biomedical and Behavioral Research.* US Government Printing Office, Washington DC. Available at: http://www.hhs.gov/ohrp/humansubjects/guidance/belmont.htm [Accessed 30/03/08].

National Bioethics Advisory Commission (1998). *Research Involving Persons with Mental Disorders That May Affect Decisionmaking Capacity: Volume 1 – Report and Recommendations of the National Bioethics Advisory Commission.* National Bioethics Advisory Commission, Rockville, Maryland. Available at: http://bioethics.georgetown.edu/nbac/capacity/TOC.htm [Accessed 30/03/08].

National Institutes of Health (1998). *NIH Policy and Guidelines on the Inclusion of Children as Participants in Research Involving Human Subjects.* National Institutes of Health, Rockville, Maryland. Available at: http://grants1.nih.gov/grants/guide/notice-files/not98-024.html. [Accessed 30/03/08].

Neugroschl, J. and Davis, K. L. (2002). 'Biological Markers in Alzheimer Disease'. *American Journal of Geriatric Psychiatry*, vol. 10, pp. 660–677.

Neves, M. D. C. P. (2007). 'The New Vulnerabilities Raised by Biomedical Research'. In: M. Häyry, T. Takala and P. Herissone-Kelly (eds.) *Ethics in Biomedical Research: International Perspectives*, Rodopi, Amsterdam and New York, pp. 181–192.

Nicholson, R. H. (1986). *Medical Research with Children: Ethics, Law and Practice*, Oxford University Press, Oxford.

Northoff, G. (2006). 'Neuroscience of Decision-Making and Informed Consent: An Investigation in Neuroethics'. *Journal of Medical Ethics*, vol. 32, pp. 70–73.

Nozick, R. (1974). *Anarchy, State and Utopia*, Blackwell, Oxford.

Nuffield Council on Bioethics (1998). *Mental Disorders and Genetics: The Ethical Context.* Nuffield Council on Bioethics, London. Available at: http://www.nuffieldbioethics.org/fileLibrary/pdf/mentaldisorders2001.pdf [Accessed 30/03/08].

The Nuremberg Code (1947). Available from National Institutes of Health Office of Human Subjects Research: http://ohsr.od.nih.gov/guidelines/nuremberg.html [Accessed 30/03/08].

Nussbaum, M. C. (2001). *Upheavals of Thought: The Intelligence of the Emotions*, Cambridge University Press, New York.

Olson, E. J. (1997). *The Human Animal: Personal Identity Without Psychology*, Oxford University Press, New York.

Ondrusek, N, Abramovitch, R., Pencharz, P. and Koren, G. (1998). 'Empirical Examination of the Ability of Children to Consent to Medical Research'. *Journal of Medical Ethics*, vol. 24, pp. 158–165.

O'Neill, O. (2000). *Bounds of Justice*, Cambridge University Press, Cambridge.

O'Neill, O. (2002). *Autonomy and Trust in Bioethics*, Cambridge University Press, Cambridge.

Owen, A. M., Coleman, M. R., Boly, M., Davis, M. H., Laureys, S. L. and Pickard, J. D. (2006). 'Detecting Awareness in the Vegetative State'. *Science*, vol. 313, p. 1402.

Palmer B. W. and Jeste, D. V. (2006). 'Relationship of Individual Cognitive Abilities to Specific Components of Decisional Capacity Among Middle-Aged and Older Patients with Schizophrenia'. *Schizophrenia Bulletin*, vol. 32 (1) pp. 98–106.

Pappworth, M. H. (1967). *Human Guinea Pigs: Experimentation on Man*, Routledge & Kegan Paul, London.

Parfit, D. (1971). 'Personal Identity'. *Philosophical Review*, vol. 80, pp. 3–27.

Parfit, D. (1984). *Reasons and Persons*, Clarendon, Oxford.

Partners in the BIOMED-II Project: 'Basic Ethical Principles in Bioethics and Biolaw' (1998). *The Barcelona Declaration: Policy Proposals to the European Commission* (November 1998). Danish Centre for Ethics and Law in Nature and Society, Copenhagen. Available at: http://www.ethiclaw.dk/publication/THE%20BARCELONA%20Dec%20Enelsk.pdf [Accessed 30/03/08].

Pascal, B. (1995) [1670]. *Pensées* (translated with an introduction by A. J. Krailsheimer), Penguin, Harmondsworth.

Pattinson, S. D. (2002). *Influencing Traits Before Birth*, Ashgate, Aldershot.

Pattinson, S. D. (2006). *Medical Law and Ethics*, Thomson Sweet & Maxwell, London.

Pellegrino, E. D. (1991). 'Informal Judgments of Competence and Incompetence'. In: M. A. G. Cutter and E. E. Shelp (eds.) *Competency: A Study of Informal Competency Determinations in Primary Health Care*, Kluwer Academic, Dordrecht, pp. 29–45.

Pepper-Smith, R., Harvey, W. and Silberfeld, M. (1996). 'Competency and Practical Judgment'. *Theoretical Medicine and Bioethics*, vol. 17(2), pp. 135–150.

Percival, T. (1803). *Medical Ethics, or, A Code of Institutes and Precepts, Adapted to the Professional Conduct of Physicians and Surgeons: To Which Is Added an Appendix Containing a Discourse on Hospital Duties, also Notes and Illustrations*, Russell and Johnson, Manchester.

Phillips, J. (2003). 'Schizophrenia and the Narrative Self'. In: T. Kircher and A. David (eds.) *The Self in Neuroscience and Psychiatry*, Cambridge University Press, Cambridge, pp. 319–335.

Phillips, D. L. (1986). *Toward a Just Social Order*, Princeton University Press, Princeton, NJ.

Piaget, J. (2001) [1950]. *The Psychology of Intelligence*, Routledge, London.

Plant, R. (1991). *Modern Political Thought*, Blackwell, Oxford.

Pluhar, E. B. (1995). *Beyond Prejudice: The Moral Significance of Human and Nonhuman Animals*, Duke University Press, Durham, NC.

Post, S. G. (1995). 'Alzheimer's Disease and the "Then" Self'. *Kennedy Institute of Ethics Journal*, vol. 5(4), pp. 307–21.

Poythress, N. G. (2002). 'Obtaining Informed Consent for Research: A Model for Use with Participants Who Are Mentally Ill'. *Journal of Law, Medicine and Ethics*, vol. 30, pp. 367–374.

Poythress, N. G., Bonnie, R. J., Monahan, J., Otto, R. and Hoge, S. K. (2002). *Adjudicative Competence: The MacArthur Studies*, Kluwer, Dordrecht.

President's Commission for the Study of Ethical Problems in Medicine and Biomedical and Behavioural Research (1982). *Making Health Care Decisions: The Ethical and Legal Implications of Informed Consent in the Patient-Practitioner Relationship, Volume 1: Report*. US Government Printing Office, Washington, DC. Available at: http://www.bioethics.gov/reports/past_commissions/making_health_care_decisions.pdf [Accessed 30/03/08].

Putnam, H. (1979). 'Other Minds'. In: H. Putnam (ed.) *Mind, Language and Reality: Philosophical Papers, vol. 2*, Cambridge University Press, Cambridge, pp. 342–361.

Radden, J. (1996). *Divided Minds and Successive Selves: Ethical Issues in Disorders of Identity and Personality*, MIT, Cambridge, Mass.

Rapoport, J. L. (eds.) (2000). *Childhood Onset of "Adult" Psychopathology: Clinical and research advances*, American Psychopathological Association and American Psychiatric Press, Washington, DC.

Rasmussen v Fleming (1987). 154 Ariz. 207, 216, 741 P.2d 674.

Rawls, J. (1972). *A Theory of Justice*, Oxford University Press, Oxford.

Rawls, J. (1993). *Political Liberalism*, Columbia University Press, New York.

Rawls, J. (1999). *A Theory of Justice*, revised edition, Oxford University Press, Oxford.

Raz, J. (1986). *The Morality of Freedom*, Clarendon, Oxford.

Re AK (Medical Treatment: Consent) [2001] 1 FLR 129.

Re B (adult: refusal of medical treatment) [2002] EWHC 429 (Fam).

Re C (adult: refusal of medical; treatment) [1994] 1 All ER 819.

Re MB (medical treatment) (1997) 38 BMLR 175 (CA).

Re R (A Minor) (Wardship: Consent to Treatment) [1992] Fam. 11.

Re T (adult: refusal of treatment) [1992] 4 All ER 349.

Re W (A Minor) (Medical Treatment: Court's Jurisdiction) [1993] Fam. 64.

Read, R. (2001). 'On Approaching Schizophrenia Through Wittgenstein'. *Philosophical Psychology*, vol. 14(4), pp. 449–475.

Rees, D. and Rose, S. (eds.) (2004). *The New Brain Sciences: Perils and Prospects*, Cambridge University Press, Cambridge.

Regis Jr., E. (ed.) (1984). *Gewirth's Ethical Rationalism: Critical Essays with a Reply by Alan Gewirth*, The University of Chicago Press, Chicago, IL.

Reich, W. T. (1978). 'Ethical Issues Related to Research Involving Elderly Subjects'. *The Gerontologist,* vol. 18(4), pp. 326–335.

Reiss, A. L., Abrams, M. T., Singer, H. S., Ross, J. L. and Denckla, M. B. (1996). 'Brain Development, Gender and IQ in Children: A Volumetric Imaging Study'. *Brain*, vol. 119, pp. 1763–1774.

Resnik, D. B. (2004). 'The Distribution of Biomedical Research Resources and International Justice'. *Developing World Bioethics*, vol. 4(1), pp. 42–57.

Reynolds, C. F. (2002). 'Advancing Research in Decision-Making Capacity'. *American Journal of Geriatric Psychiatry*, vol. 10(2), pp. 117–119.

Reznek, L. (1991). *The Philosophical Defence of Psychiatry*, Routledge, London.

Richter, D. (ed.) (1984). *Research in Mental Illness*, Heinemann Medical Books, London.

Riese v St. Mary's Hospital 243 Cal. Rptr 241 (1987).

Rikkert, M. G., van den Bercken, J. H., ten Have, H. A. and Hoefnagels, W. H. (1997). 'Experienced consent in geriatrics research: a new method to optimize the capacity to consent in frail elderly subjects'. *Journal of Medical Ethics*, vol. 23, pp. 271–276.

Rizzolatti, G. and Sinigaglia, C. (trans. F. Anderson) (2007). *Mirrors in the Brain: How Our Minds Share Actions and Emotions*, Oxford University Press, Oxford.

Roberts, L. W. (2000). 'Evidence-Based Ethics and Informed Consent in Mental Illness Research'. *Archives of General Psychiatry*, vol. 57(6), pp 540–542.

Roberts, L W. (2002). 'Informed Consent and the Capacity for Voluntarism'. *American Journal of Psychiatry*, vol. 159, pp. 705–712.

Roberts, L. W., Warner T. D. and Brody, J. L. (2000). 'Perspectives of Patients with Schizophrenia and Psychiatrists Regarding Ethically Important Aspects of Research Participation'. *American Journal of Psychiatry*, vol. 157, pp. 67–74.

Robertson, J. A. (1991). 'The Geography of Competence'. In: M. A. G. Cutter and E. E. Shelp (eds.) *Competency: A Study of Informal Competency Determinations in Primary Care*, Kluwer Academic, Dordrecht, pp. 127–148.

Ross, L. F. (1998). *Children, Families and Health Care Decision Making*, Oxford University Press, Oxford.

Ross, L. F. (2006). *Children in Medical Research: Access versus Protection*, Oxford University Press, Oxford.

Roth, L. H. (1979). 'A Commitment Law for Patients, Doctors and Lawyers'. *American Journal of Psychiatry*, vol. 136, pp. 1121–1127.

Roth, L. H., Meisel, A., Lidz, C. W. and Appelbaum, P. S. (1977). 'Tests of Competency to Consent to Treatment'. *American Journal of Psychiatry*, vol. 134, pp. 279–284

Roth, L., Meisel, A. and Lidz, C. W. (1982). 'The Dilemma of Denial in the Assessment of Competence to Refuse Treatment'. *American Journal of Psychiatry*, vol. 139, pp. 910–913.

Roth, L. H., Lidz, C. W., Meisel, A., Soloff, P. H., Kaufmann, K., Spiker, D. G., Foster, F. G. (1982). 'Competency to Decide About Treatment or Research: An Overview of some Empirical Data'. *International Journal of Law and Psychiatry*, vol. 5, pp. 29–50.

Rudnick, A. (2002). 'Depression and Competence to Refuse Psychiatric Treatment'. *Journal of Medical Ethics*, vol. 28(3), pp. 151–155.
Ryle, G. (1949). *The Concept of Mind*, Hutchinson, London.
Sadler, J. Z. (2005). *Values and Psychiatric Diagnosis*, Oxford University Press, Oxford.
Saks, E. R. and Jeste, D. V. (2006). 'Capacity to Consent to or Refuse Treatment and/or Research: Theoretical Considerations'. *Behavioral Sciences and the Law*, vol. 24(4), pp. 411–429.
Sandel, M. J. (1998). *Liberalism and the Limits of Justice*, second edition, Cambridge University Press, Cambridge.
Sass, L. A. (1995). *The Paradoxes of Delusion: Wittgenstein, Schreber and the Schizophrenic Mind*, Cornell University Press, Ithaca, New York.
Schaffner, E. L. (1991). 'Competency: A Triaxial Concept'. In: M. A. G. Cutter and E. E. Shelp (eds.) *Competency: A Study of Informal Competency Determinations in Primary Care*, Kluwer Academic, Dordrecht, pp. 253–281.
Schneider, P. L. and Bramstedt, K. A. (2006). 'When Psychiatry and Bioethics Disagree About Patient Decision Making Capacity (DMC)'. *Journal of Medical Ethics*, vol. 32, pp. 90–93.
Schloendorff v Society of New York Hospital (1914). 211 NY 125.
Shanteau, J., Johnson, P. and Smith, K. (eds.) (2004). *Psychological Investigations of Competence in Decision Making*, Cambridge University Press, Cambridge.
Shapiro, M. H. (1974). 'Legislating the Control of Behavior Control: Autonomy and the Coercive Use of Organic Therapies'. *Southern California Law Review*, vol. 47, pp. 237–356.
Shiffrin, S. V. (2004). 'Autonomy, Beneficence and the Permanently Demented'. In: J. Burley (ed.) *Dworkin and His Critics*, Oxford: Blackwell, pp. 195–217.
Shildrick, M. (1997). *Leaky Bodies and Boundaries: Feminism, Postmodernism and (Bio)ethics*, Routledge, London.
Shoemaker, D. (2007). 'Personal Identity and Ethics'. In: E. N. Zalta (ed.) *The Stanford Encyclopedia of Philosophy (Spring 2007 Edition)*. Available at: http://plato.stanford.edu/archives/spr2007/entries/identity-ethics [Accessed: 30/03/08].
Silberfeld, M. and Checkland, D. (1999). 'Faulty Judgment, Expert Opinion and Decision-Making Capacity'. *Theoretical Medicine and Bioethics*, vol. 20, pp. 377–393.
Silberfeld, M. (2002). 'Vulnerable Persons'. In: D. C. Thomasma, D. N. Weisstub and C. Hervé (eds.) *Personhood and Health Care*, Kluwer, Dordrecht, pp. 299–317.
Silver, M. (2002). 'Reflections on Determining Competency'. *Bioethics*, vol. 16(5), pp. 455–468.
Simard, M. and van Reekum, R. (1999). 'Memory Assessment in Studies of Cognition-Enhancing Drugs for Alzheimer's Disease'. *Drugs and Aging*, vol. 14(3), pp. 197–230.
Simpson, J. and Weiner, E. (eds.) (1989). *The Oxford English Dictionary* (in twenty volumes), second edition, Clarendon, Oxford.
Singer, P. (1995). *Animal Liberation*, second edition, Pimlico, London.
Silverman, H. (1997). 'The Role of Emotions in Decisional Competence, Standards of Competency and Altruistic Acts'. *Journal of Clinical Ethics*, vol. 8(2), pp. 171–175.
Skene, L. (1991). 'Risk-Related Standard Inevitable in Assessing Competence'. *Bioethics*, vol. 5 (2), pp. 113–117.
Spaak, T. (1994). *The Concept of Legal Competence: An Essay in Conceptual Analysis*, Dartmouth, Aldershot.
Spaak, T. (2003). 'Norms That Confer Competence'. *Ratio Juris*, vol. 16(1), pp. 89–104.
Spence, E. H. (2006). *Ethics Within Reason: A Neo-Gewirthian Approach*, Lexington Books, Lanham, MD.
Spicker, P. (1990). 'Mental Handicap and Citizenship'. *Journal of Applied Philosophy*, vol. 7(2), pp. 139–151.
State of Tennessee Department of Human Services v Mary C. Northern (1978). 563 S.W. 2d 197 (Tenn Ct App).
Stefan, S. (1993). 'Silencing the Different Voice: Competence, Feminist Theory and the Law'. *University of Miami Law Review*, vol. 47 pp. 763–815.
Steigleder, K. (1998). 'The Moral Status of Potential Persons'. In: E. Hildt and D. Mieth (eds.) *In Vitro Fertilisation in the 1990s: Towards a Medical, Social and Ethical Evaluation*, Ashgate, Aldershot, pp. 239–246.

Sternberg, R. J. and Kolligian Jr., J. (eds.) (1990). *Competence Considered*, Yale University Press, New Haven, CT.

Stiles, P. G., Poythress, N. G., Hall, A., Falkenbach, D. and Williams, R. (2001). 'Improving Understanding of Research Consent Disclosures Among Persons With Mental Illness'. *Psychiatric Services*, vol. 52(6), pp. 780–785.

Strawson, G. (1997). 'The Self'. *Journal of Consciousness Studies*, vol. 4(5–6), pp. 405–428.

Sullivan, M. D. and Youngner, S. J. (1994). 'Depression, Competence and the Right to Refuse Lifesaving Medical Treatment'. *American Journal of Psychiatry*, vol. 151, pp. 971–978.

Sumner, L. W. (1987). *The Moral Foundations of Rights*, Clarendon, Oxford.

Szasz, T. (1974). *Law, Liberty and Psychiatry: An Inquiry into the Social Use of Mental Health Practices*, Routledge & Kegan Paul, London.

Szasz, T. (2003) [1974]. *The Myth of Mental Illness: Foundations of a Theory of Personal Conduct*, revised edition, Perennial, New York.

Tait, A. R., Voepel-Lewis, T. and Malviya, S. (2003). 'Do they understand? (Part II)': Assent of Children Participating in Clinical Anaesthesia and Surgery Research'. *Anaesthesiology*, vol. 98, pp. 609–614.

Tan, J. O. A. and Fegert, J. M. (2004). 'Capacity and Competence in Child and Adolescent Psychiatry'. *Health Care Analysis*, vol. 12(4), pp. 285–294.

Tan, J. O. A. and McMillan, J. R. (2004). 'The Discrepancy Between the Legal Definition of Capacity and the British Medical Association's Guidelines'. *Journal of Medical Ethics*, vol. 30, pp. 427–429.

Tan, J. A. O., Stewart, A., Fitzpatrick, R. and Hope, T. (2006). 'Competence to Make Treatment Decisions in Anorexia Nervosa: Thinking Processes and Values'. *Philosophy, Psychiatry and Psychology*, vol. 13(4), pp. 267–282.

Tancredi, L. (1982). 'Competency for Informed Consent: Conceptual limits of Empirical Data'. *International Journal of Law and Psychiatry*, vol. 5, pp. 51–63.

Taylor, C. (1985a). 'What is Human Agency?' In: C. Taylor (ed.) *Philosophical Papers, Volume 1: Human Agency and Language*, Cambridge University Press, Cambridge, pp. 15–44.

Taylor, C. (1985b). 'The Concept of a Person'. In: C. Taylor (ed.) *Philosophical Papers, Volume 1: Human Agency and Language*, Cambridge University Press, Cambridge, pp. 97–114.

The Medicines for Human Use (Clinical Trials) Regulations (2004). SI 2004/1031, TSO, London. Available at: http://www.opsi.gov.uk/si/si2004/20041031.htm [Accessed 30/03/08].

The Mental Capacity Act 2005 (Independent Mental Capacity Advocates) (Expansion of Role) Regulations (2006). SI 2006/2883. TSO, London. Available at: http://www.opsi.gov.uk/SI/si2006/uksi_20062883_en.pdf [Accessed 30/03/08].

Thomas, J. E. and Waluchow, W. J. (1998). *Well and Good: Case Studies in Biomedical Ethics*, third edition, Broadview, Ontario.

Thomasma, D. C. and Pellegrino, E. D. (1998). 'Preface: Medicine, Science, Self-interest: Value Sets in Conflict in Human Experimentation. In: D. N. Weisstub (ed.) *Research on Human Subjects: Ethics, Law and Social Policy*, Pergamon, Oxford, pp. xvii–xxvi.

Thompson, J. J. (1990). *The Realm of Rights*, Harvard University Press, Cambridge, MA.

Thor v Superior Court 855 P.2d 375 (Cal. 1993).

Thornicroft, G. (2006). *Shunned: Discrimination against People with Mental Illness*, Oxford University Press, Oxford.

Toddington, S. (1993). *Rationality, Social Action and Moral Judgment,* Edinburgh University Press, Edinburgh.

Toner, K. and Schwartz, R. (2003). 'Why a Teenager over Age 14 Should Be Able to Consent Rather than Merely Assent to Participation as a Human Subject of Research'. *American Journal of Bioethics*, vol. 3(4), pp. 38–40.

Tremain, S. (2002). 'On the Subject of Impairment'. In: T. Shakespeare and M. Corker (eds.) *Disability/Postmodernity: Embodying Disability Theory*, Continuum, London, pp. 32–54.

UK Department of Health (2001a). *Reference Guide to Consent for Examination or Treatment*. Department of Health, London. Available at: http://www.dh.gov.uk/prod_consum_dh/groups/dh_digitalassets/@dh/@en/documents/digitalasset/dh_4019079.pdf [Accessed 30/03/08].

UK Department of Health (2001b). *Good Practice in Consent Implementation Guide: Consent to Examination or Treatment*. Department of Health, London. Available at: http://www.dh.gov.uk/prod_consum_dh/groups/dh_digitalassets/@dh/@en/documents/digitalasset/dh_4019061.pdf [Accessed 30/03/08].

UK Department of Health (2005). *Consultation on the Independent Mental Capacity Advocate Service*. Department of Health, London. Available at: http://www.dh.gov.uk/en/Consultations/Closedconsultations/DH_4119900 [Accessed: 30/03/08].

UK Department of Health (2006). *Consultation on the Independent Mental Capacity Advocate (IMCA) Service 2005: Summary of responses*. Department of Health, London. Available at: http://www.dh.gov.uk/prod_consum_dh/groups/dh_digitalassets/@dh/@en/documents/digitalasset/dh_4133915.pdf [Accessed 30/03/08].

Unger, R. (1984). 'A Programme for Late Twentieth Century Psychiatry'. *American Journal of Psychiatry*, vol. 139(2) pp. 155–164.

United Nations General Assembly (2006). *Convention on the Rights of Persons with Disabilities*. United Nations, New York. Available at: http://daccessdds.un.org/doc/UNDOC/LTD/N06/645/30/PDF/N0664530.pdf?OpenElement [Accessed 30/03/08].

US v Charters (1987). 829 F.2d 479.

Vallentyne, P. (1991). 'Gauthier's Three Projects'. In: P. Vallentyne (ed.) *Contractarianism and Rational Choice: Essays on David Gauthier's Morals by Agreement*, Cambridge University Press, New York, pp. 1–11.

Vanderpool, H. Y. (1991). 'The Competency of Definitions of Competency'. In: M. A. G. Cutter and E. E. Shelp (eds.) *Competency: A Study of Informal Competency Determinations in Primary Care*, Kluwer Academic, Dordrecht, pp. 197–209.

Vaswani, N. A. (2000). *Competence as a Condition for Informed Consent in Medical Contexts*. Unpublished M. Phil. thesis, King's College London.

Veatch, R. M. (1987). *The Patient as Partner: A Theory of Human Experimentation Ethics*, Indiana University Press, Bloomington, IN.

Wahlström, J. (ed.) (1998). *Genetics and Psychiatric Disorders (Wenner-Gren International Series vol. 69)*, Pergamon, Oxford.

Waldron, J. (1993). *Liberal Rights: Collected Papers 1981–1991*. Cambridge University Press, Cambridge.

Warner, R. (2003). *Recovery from Schizophrenia: Psychiatry and Political Economy*, third edition, Routledge, London.

Warren, M. A. (1997). *Moral Status: Obligations to Persons and Other Living Things*, Oxford University Press, Oxford.

Washington v Harper (1990). 494 U.S. 210, 110 S.Ct. 1028.

Wear, S. (1998). *Informed Consent: Patient Autonomy and Clinician Beneficence within Health Care*, second edition, Georgetown University Press, Washington, DC.

Weithorn, L. A. and Campbell, S. B. (1982). 'The Competency of Children and Adolescents to Make Informed Consent Treatment Decisions'. *Child Development*, vol. 53, pp. 1589–1598.

Weithorn, L. A. and Schreber, D. G. (1994). 'Children's Involvement in Research Participation Decisions: Psychological Considerations'. In: M. A. Grodin and L. A. Glantz (eds.) *Children as Research Subjects: Science, Ethics and Law*, Oxford University Press, New York, pp. 133–179.

Weisstub, D. N. and Arboleda-Florez, J. (1998). 'Ethical Research with Vulnerable Populations: The Developmentally Disabled'. In: D. N. Weisstub (ed.) *Research on Human Subjects: Ethics, Law and Social Policy*, Pergamon, Oxford, pp. 479–494.

Weisstub, D. N. and Thomasma, D. C. (2002). 'Human Dignity, Vulnerability, Personhood'. In: D. C. Thomasma, D. N. Weisstub and C. Hervé (eds.) *Personhood and Health Care*, Kluwer, Dordrecht, pp. 317–332.

Weisstub, D. N., Verdun-Jones, S. N. and Walker, J. (1998a). 'Biomedical Experimentation with Children: Balancing the Need for Protective Measures with the Need to Respect Children's Developing Ability to Make Significant Life Decisions for Themselves'. In: D. N. Weisstub (ed.) *Research on Human Subjects: Ethics, Law and Social Policy*, Pergamon, Oxford, pp. 380–404.

Weisstub, D. N., Verdun-Jones, S. N. and Walker, J. (1998b). 'Biomedical Experimentation Involving Elderly Subjects: The Need to Balance Limited, Benevolent Protection with Recognition of a Long History of Autonomous Decision-Making'. In: D. N. Weisstub (ed.) *Research on Human Subjects: Ethics, Law and Social Policy*, Pergamon, Oxford, pp. 405–432.

Welie, J. V. M. and Welie, S. P. K. (2001). 'Patient Decision Making Competence: Outlines of a Conceptual Analysis'. *Medicine, Health Care and Philosophy*, vol. 4(2), pp. 127–138.

Welie, S. P. K. and Berghmans, R. L. (2006). 'Inclusion of Patients with Severe Mental Illness in Clinical Trials: Issues and Recommendations Surrounding Informed Consent'. *CNS Drugs*, vol. 20(1), pp. 67–83.

Wellman, C. (1995). *Real Rights*, Oxford University Press, Oxford.

Wellman, C. H. (1999). *The Proliferation of Rights: Moral Progress or Empty Rhetoric?* Westview, Boulder, CO.

Wellman, H. M. (1992). *The Child's Theory of Mind*, MIT, Cambridge, MA.

Wendler, D. and Shah. S. (2003). 'Should Children Decide Whether They Are Enrolled in Non-Beneficial Research?' *American Journal of Bioethics*, vol. 3(4), pp. 1–7.

Wertheimer, A. (2003). *Consent to Sexual Relations*, Cambridge University Press, Cambridge.

White, B. C. (1994). *Competence to Consent*. Georgetown University Press, Washington, DC.

White, P. D. and Denise, S. H. (1991). 'Medical Treatment in the Eyes of the Law'. In: M. A. G. Cutter and E. E. Shelp (eds.) *Competency: A Study of Informal Competency Determinations in Primary Care*, Kluwer Academic, Dordrecht, pp. 149–164.

Wicclair, M. R. (1991a). 'Patient Decision-Making Capacity and Risk'. *Bioethics*, vol. 5(2), pp. 87–104.

Wicclair, M. R. (1991b). 'A Response to Brock and Skene'. *Bioethics*, vol. 5(2), pp. 118–122.

Wicclair, M. R. (1993). *Ethics and the Elderly*, Oxford University Press, New York.

Wicclair, M. R. (1999). 'The Continuing Debate Over Risk-Related Standards of Competence'. *Bioethics*, vol. 13(2), pp. 149–153.

Wikler, D. (1979). 'Paternalism and the Mildly Retarded'. *Philosophy and Public Affairs,* vol. 8(4), pp. 377–392.

Wilkes, K. V. (1988). *Real People: Personal Identity without Thought Experiments*, Clarendon, Oxford.

Wilkinson, H. (2002). 'Including People with Dementia in Research: Methods and Motivations'. In: H. Wilkinson (ed.) *The Perspectives of People with Dementia: Research Methods and Motivations*, Jessica Kingsley, London, pp. 9–24.

Wilks, I. (1997). 'The Debate over Risk-Related Standards of Competence'. *Bioethics*, vol. 11(5), pp. 413–426.

Wilks, I. (1999). 'Asymmetrical Competence'. *Bioethics*, vol. 13(2), pp. 154–159.

Williams, B. (1970). 'The Self and the Future'. *The Philosophical Review*, vol. 79, pp. 161–180.

Williams, B. (1973). 'Personal Identity and Individuation'. In: B. Williams (ed.) *Problems of the Self: Philosophical Papers 1956–1972*. Cambridge University Press, Cambridge, pp. 1–18.

Williams, J. B. W. and Ell, K. (eds.) (1999). *Advances in Mental Health Research: Implications for Practice*, NASW, Washington, DC.

Winick, B. J. (1995). 'The Side Effects of Incompetency Labelling and the Implications for Mental Health Law'. *Psychology, Public Policy and Law*, vol. 1(1), pp. 6–42.

Wirshing, D. A., Wirshing, W. C., Marder, S. R., Liberman, R. P. and Mintz, J. (1998). 'Informed Consent: Assessment of Comprehension'. *American Journal of Psychiatry*, vol. 155, pp. 1508–1511.

Wittchen, H. U., Jacobi, F. (2005). 'Size and Burden of Mental Disorders in Europe - a Critical Review and Appraisal of 27 Studies'. *European Neuropsychopharmacology*, vol. 15, pp. 357–376.

Wittgenstein, L. (2002) [1953]. *Philosophical Investigations* (trans. G. E. M. Anscombe). Routledge, London.

Wolf, S. M. (ed.) (1996). *Feminism and Bioethics: Beyond Reproduction*, Oxford University Press, New York.

Wolpert, L. (2001). *Malignant Sadness: The Anatomy of Depression*, Faber & Faber, London.

Wong, J. G., Clare, I. C. H., Holland, A. J., Watson, P. C. and Gunn, M. J. (2000). 'The Capacity of People with a 'Mental Disability' to Make a Health Care Decision'. *Psychological Medicine*, vol. 30, pp. 295–306.

Wood, A. W. (1999). *Kant's Ethical Thought*, Cambridge University Press, Cambridge.

World Health Organisation (1992). *The ICD-10 Classification of Mental and Behavioural Disorders: Clinical Descriptions and Diagnostic Guidelines*, World Health Organisation, Geneva.

World Medical Association (2004). *Declaration of Helsinki: Ethical Principles for Medical Research Involving Human Subjects*, adopted by 18th WMA (1964), revised by 52nd WMA (2000), with note of clarification on Paragraph 30 added by the WMA General Assembly, Tokyo, 2004. Available at: http://www.wma.net/e/policy/pdf/17c.pdf [Accessed 30/03/08].

Wrigley, A. (2007). 'Proxy Consent: Moral Authority Misconceived'. *Journal of Medical Ethics*, vol. 33, pp. 527–531.

Yarborough, M. (2002). 'Adults Are Not Big Children: Examining Surrogate Consent to Research Using Adults with Dementia'. *Cambridge Quarterly of Healthcare Ethics*, vol. 11, pp. 160–168.

Zaubler, T. S., Viederman, M. and Fins, J. J. (1996). 'Ethical, Legal and Psychiatric Issues in Capacity, Competency and Informed Consent: An Annotated Bibliography'. *General Hospital Psychiatry*, vol. 18, pp. 155–172.

Zerr I. and Poser S. (2002). 'Clinical Diagnosis and Differential Diagnosis of CJD and vCJD'. *APMIS (Acta Pathologica, Microbiologica et Immunologica Scandinavica)*, vol. 110, pp. 88–98.

Zoboli, E. L. C. P. (2007). 'Vulnerability in Biomedical Research: A Framework for Analysis'. In: M. Häyry, T. Takala and P. Herissone-Kelly (eds.) *Ethics in Biomedical Research: International Perspectives*, Rodopi, Amsterdam and New York, pp. 167–179.

Kolibri 360 GmbH
Ferdinand-Jühlke-Straße 7
99095 Erfurt, Deutschland
produktsicherheit@kolibri360.de